T0220555

Lecture Notes in Mathematics

Edited by J.-M. Morel, F. Takens and B. Teissier

Editorial Policy
for the publication of monographs

1. Lecture Notes aim to report new developments in all areas of mathematics – quickly, informally and at a high level. Monograph manuscripts should be reasonably self-contained and rounded off. Thus they may, and often will, present not only results of the author but also related work by other people. They may be based on specialized lecture courses. Furthermore, the manuscripts should provide sufficient motivation, examples and applications. This clearly distinguishes Lecture Notes from journal articles or technical reports which normally are very concise. Articles intended for a journal but too long to be accepted by most journals, usually do not have this "lecture notes" character. For similar reasons it is unusual for doctoral theses to be accepted for the Lecture Notes series.

2. Manuscripts should be submitted (preferably in duplicate) either to one of the series editors or to Springer-Verlag, Heidelberg. In general, manuscripts will be sent out to 2 external referees for evaluation. If a decision cannot yet be reached on the basis of the first 2 reports, further referees may be contacted: the author will be informed of this. A final decision to publish can be made only on the basis of the complete manuscript, however a refereeing process leading to a preliminary decision can be based on a pre-final or incomplete manuscript. The strict minimum amount of material that will be considered should include a detailed outline describing the planned contents of each chapter, a bibliography and several sample chapters.
Authors should be aware that incomplete or insufficiently close to final manuscripts almost always result in longer refereeing times and nevertheless unclear referees' recommendations, making further refereeing of a final draft necessary.
Authors should also be aware that parallel submission of their manuscript to another publisher while under consideration for LNM will in general lead to immediate rejection.

3. Manuscripts should in general be submitted in English.
Final manuscripts should contain at least 100 pages of mathematical text and should include
– a table of contents;
– an informative introduction, with adequate motivation and perhaps some
 historical remarks: it should be accessible to a reader not intimately familiar
 with the topic treated;
– a subject index: as a rule this is genuinely helpful for the reader.

Continued on inside back-cover

Lecture Notes in Mathematics 1757

Editors:
J.-M. Morel, Cachan
F. Takens, Groningen
B. Teissier, Paris

Springer
Berlin
Heidelberg
New York
Barcelona
Hong Kong
London
Milan
Paris
Singapore
Tokyo

Robert R. Phelps

Lectures on Choquet's Theorem

Second Edition

Springer

Author

Robert R. Phelps
Department of Mathematics
Box 354350
University of Washington
Seattle WA 98195, USA

E-mail: phelps@math.washington.edu

Cataloging-in-Publication Data applied for

Die Deutsche Bibliothek - CIP-Einheitsaufnahme

Phelps, Robert R.:
Lectures on Choquet's theorem / Robert R. Phelps. - 2. ed.. - Berlin ;
Heidelberg ; New York ; Barcelona ; Hong Kong ; London ; Milan ; Pan.
; Singapore ; Tokyo : Springer, 2001
  (Lecture notes in mathematics ; 1757)
  ISBN 3-540-41834-2

The first edition was published by Van Nostrand, Princeton, N.J. in 1966

Mathematics Subject Classification (2000): 46XX

ISSN 0075-8434
ISBN 3-540-41834-2 Springer-Verlag Berlin Heidelberg New York

Springer-Verlag Berlin Heidelberg New York
a member of BertelsmannSpringer Science+Business Media GmbH

http://www.springer.de

© Springer-Verlag Berlin Heidelberg 2001
Printed in Germany

Typesetting: Camera-ready TEX output by the authors
SPIN: 10759944   41/3142-543210 - Printed on acid-free paper

Preface to Second Edition

On a delightful Belgian canal trip during a break from a Mons University conference in the summer of 1997, Ward Henson suggested that I make available a LaTeX version of this monograph, which was originally published by Van Nostrand in 1966 and has long been out of print. Ms. Mary Sheetz in the University of Washington Mathematics Department office expertly and quickly carried out the difficult job of turning the original text into a LaTeX file, providing the foundation for this somewhat revised and expanded version. I am delighted that it is being published by Springer–Verlag.

Since 1966 there has been a great deal of research related to Choquet's theorem, and there was considerable temptation to include much of it, easily doubling the size of the original volume. I decided against doing so for two reasons. First, there exist readable treatments of most of this newer material. Second, the feedback I have received over the years has indicated that the small size of the first edition made it an easily accessible introduction to the subject, suitable for a one–term seminar (of the type which generated it in the first place). This edition does include some newer results which are closely related to the original text, but some other more recent material is merely summarized in the final section. It also incorporates a number of suggestions and corrections to the first edition which I have received over the years. I thank all those who have helped me in this regard, especially Robert Burckel and Christian Skau (who have surely forgotten the letters they sent me in the 70's) as well as my colleague Isaac Namioka. Of course, I'm the one responsible for any new errors. I am grateful to Elaine Phelps, who tolerated my preoccupation with this task (during both editions); her support made the work easier.

R. R. P.

Seattle, Washington
December, 2000

Preface to First Edition

These notes are a revised and expanded version of mimeographed notes originally prepared for a seminar during Spring Quarter, 1963, at the University of Washington. They are designed to be read by anyone with a knowledge of the Krein-Milman theorem and the Riesz representation theorem (along with the functional analysis and measure theory implicit in an understanding of these theorems). The only major theorem which is used without proof is the one on "disintegration of measures" in Section 15.

The author is indebted to many people who helped, directly or indirectly, in the preparation of these notes. He has especially benefited from the Walker-Ames lectures at the University of Washington in the summer of 1964, by Professor G. Choquet, and from the stay at the same institution during 1963 by Professor P. A. Meyer. He has received helpful comments from many of his colleagues, as well as from Professors N. Rothman and A. Peressini, who used the earlier version in a seminar at the University of Illinois. Finally, he wishes to thank Professor J. Feldman for permitting the inclusion of the unpublished material in Section 12 on invariant and ergodic measures.

A note to the reader: Although the applications of the theory are interspersed throughout the notes, they are never needed for subsequent material. Thus, Sections 2, 5, 7, 9 or 12, for instance, may be put aside for later reading without encountering any difficulties. (To omit them entirely, however, would cut the subject off from its many and interesting connections with other parts of mathematics.)

R. R. P.

Seattle, Washington
March, 1965

Contents

1 Introduction. The Krein-Milman theorem as an integral representation theorem

The simplest example of a theorem of the type with which we will be concerned is the following classical result of Minkowski (see the exercise on page 7).

If X is a compact convex subset of a finite-dimensional vector space E, and if x is an element of X, then x is a finite convex combination of extreme points of X. Thus, there exist extreme points $x_1, \ldots, x_k$ and positive numbers $\mu_1, \ldots, \mu_k$ with $\Sigma_1^k \mu_i = 1$ such that $x = \Sigma \mu_i x_i$. We now reformulate this representation of x as an "integral representation." For any point y of X let ε_y be the "point mass" at y, i.e., ε_y is the Borel measure which equals 1 on any Borel subset of X which contains y, and equals 0 otherwise. Abbreviating ε_{x_i} by ε_i, let $\mu = \Sigma \mu_i \varepsilon_i$; then μ is a regular Borel measure on X, $\mu \geq 0$, and $\mu(X) = 1$. Furthermore, for any continuous linear functional f on E, we have $f(x) = (\Sigma \mu_i f(x_i) =) \int_X f d\mu$. This last assertion is what we mean when we say that μ represents x.

DEFINITION. *Suppose that X is a nonempty compact subset of a locally convex space E, and that μ is a probability measure on X. (That is, μ is a nonnegative regular Borel measure on X, with $\mu(X) = 1$.) A point x in E is said to be* represented *by μ if $f(x) = \int_X f d\mu$ for every continuous linear functional f on E. (We will sometimes write $\mu(f)$ for $\int_X f d\mu$, when no confusion can result.) (Other terminology: "x is the barycenter of μ," "x is the resultant of μ.")*

The restriction that E be locally convex is simply to insure the existence of sufficiently many functionals in E^* to separate points; this guarantees that there is at most one point represented by μ. Later, we will want to consider measures on other σ-rings, but the Borel measures suffice for the present.

Note that any point x in X is trivially represented by ε_x; the interesting (and important) fact brought out by the above example is that for a compact convex subset X of a finite dimensional space, each x in X may be represented by a probability measure which is "supported" by the extreme points of X.

DEFINITION. *If μ is a nonnegative regular Borel measure on the compact Hausdorff space X and S is a Borel subset of X, we say that μ is supported by S if $\mu(X \setminus S) = 0$.*

We may now formulate the problems which concern us: *If X is a compact convex subset of a locally convex space E, and x is an element of X, does there exist a probability measure μ on X which is supported by the extreme points of X and which represents x? If μ exists, is it unique?* Choquet [17] has shown that, under the additional hypothesis that X be metrizable, the first question has an affirmative answer, while an affirmative answer to the second question depends on a certain geometrical property of X. Bishop and de Leeuw [9] have shown that if we allow more general measures than Borel measures, then the answer to the first question is affirmative (*without* additional hypotheses on X).

In the above example, the introduction of an integral in place of a convex combination was a bit artificial. It seems worthwhile to translate two well-known theorems (the Riesz representation theorem and the Krein-Milman theorem) into the language which we have introduced; in these instances the use of integrals is quite natural. It will also make clear exactly how the theorems of Choquet and Bishop-de Leeuw generalize the Krein-Milman theorem.

Let Y be a compact Hausdorff space, $C(Y)$ the Banach space of all continuous real-valued functions on Y (supremum norm), and X the set of all continuous linear functionals L on $C(Y)$ such that $L(1) = 1 = \|L\|$. Then X is a compact convex subset of the locally convex space $E = C(Y)^*$ (in its weak* topology) and the Riesz theorem asserts that to each L in X there corresponds a unique probability measure μ on Y such that $L(f) = \int_Y f \, d\mu$ for each f in $C(Y)$. By a well-known theorem [28, p. 442], Y is homeomorphic (via the natural embedding $y \to$ (evaluation at y)) with the set of extreme points of X, so we may consider μ as a probability measure on the Borel subsets of X which vanish on the open set $X \setminus Y$, and hence

μ is supported by the extreme points of X. One need only recall that E^*, the space of weak* continuous linear functionals on $C(Y)^*$, consists precisely those of all functionals of the form $L \to L(f)$ (f in $C(Y)$) in order to see that this is a representation theorem of the type we are considering.

There are two points in the above paragraph which, it should be emphasized, are *not* characteristic of the general situation. First, the extreme points of X formed a compact (hence a Borel) subset; second, the representation was unique. (We will return to these points a little later.) It is clear that any probability measure μ on Y defines (by $f \to \int_X f \, d\mu$) a linear functional on $C(Y)$ which is in X. This fact *is* true under fairly general circumstances, as the next result shows. First, recall that a function ϕ from one linear space to another is *affine* provided $\phi(\lambda x + (1 - \lambda)y) = \lambda\phi(x) + (1 - \lambda)\phi(y)$ for any x, y and any real λ.

PROPOSITION 1.1 *Suppose that Y is a compact subset of a locally convex space E, and that the closed convex hull X of Y is compact. If μ is a probability measure on Y, then there exists a unique point x in X which is represented by μ, and the function $\mu \to$ (resultant of μ) is an affine weak* continuous map from $C(Y)^*$ into X.*

PROOF. We want to show that the compact convex set X contains a point x such that $f(x) = \int_Y f \, d\mu$ for each f in E^*. For each f, let $H_f = \{y : f(y) = \mu(f)\}$; these are closed hyperplanes, and we want to show that $\cap\{H_f : f \in E^*\} \cap X$ is nonempty. Since X is compact, it suffices to show that for any finite set $f_i, \ldots, f_n$ in E^*, $\bigcap_{i=1}^{n} H_{f_i} \cap X$ is nonempty. To this end, define

$$T : E \to R^n \quad \text{by} \quad Ty = (f_1(y), f_2(y), \ldots, f_n(y));$$

then T is linear and continuous, so that TX is compact and convex. It suffices to show that $p \in TX$, where $p = (\mu(f_1), \mu(f_2), \ldots, \mu(f_n))$. If $p \notin TX$ there exists a linear functional on R^n which strictly separates p and TX; representing the functional by $a = (a_1, a_2, \ldots, a_n)$, this means that $(a, p) > \sup\{(a, Ty) : y \in X\}$. If we define g in E^* by $g = \Sigma a_i f_i$, then the last assertion becomes $\int_Y g \, d\mu > \sup g(X)$. Since $Y \subset X$ and $\mu(Y) = 1$, this is impossible, and the first part of the proof is complete. Suppose, next, that the net μ_α of probability measures on Y converges weak* in $C(X)^*$ to the probability measure μ, and let x_α and x denote their respective resultants. Since X is compact, to show that $x_\alpha \to x$ it suffices to show that every convergent subnet x_β of x_α converges to x. But if $x_\beta \to y$, say, then the corresponding subnet μ_β converges weak* to μ, and hence $f(x_\beta) = \mu_\beta(f) \to \mu(f) = f(x)$ for each f in E^*; since the latter separates points of $X, y = x$.

The hypothesis that X be compact may be avoided in those spaces E in which the closed convex hull of a compact set is always compact; for instance, if E is complete, or if E is the locally convex space obtained by taking a Banach space in its weak topology [28, p. 434].

A simple, but useful, characterization of the closed convex hull of a compact set can be given in terms of measures and their barycenters.

PROPOSITION 1.2 *Suppose that Y is a compact subset of a locally convex space E. A point x in E is in the closed convex hull X of Y if and only if there exists a probability measure μ on Y which represents x.*

PROOF. If μ is a probability measure on Y which represents x, then for each f in E^*, $f(x) = \mu(f) \leqq \sup f(Y) \leqq \sup f(X)$. Since X is closed and convex, it follows that x is in X. Conversely, if x is in X, there exists a net in the convex hull of Y which converges to x. Equivalently, there exist points y_α of the form $y_\alpha = \Sigma_{i=1}^{n_\alpha} \lambda_i^\alpha x_i^\alpha$, ($\lambda_i^\alpha > 0, \Sigma \lambda_i^\alpha = 1, x_i^\alpha$ in Y, α in some directed set) which converges to x. We may represent each y_α by the probability measure $\mu_\alpha = \Sigma \lambda_i^\alpha \varepsilon_{x_i}^\alpha$. By the Riesz theorem, the set of all probability measures on Y may be identified with a weak*-compact convex subset of $C(Y)^*$, and hence there exists a subnet μ_β of μ_α converging (in the weak* topology of $C(Y)^*$) to a probability measure μ on Y. In particular, each f in E^* is (when restricted to Y) in $C(Y)$, so $\lim f(y_\beta) = \lim \int f \, d\mu_\beta = \int f \, d\mu$. Since y_α converges to x, so does the subnet y_β, and hence $f(x) = \int_Y f \, d\mu$ for each f in E^*, which completes the proof.

The above proposition makes it easy to reformulate the Krein-Milman theorem. Recall the statement: *If X is a compact convex subset of a locally convex space, then X is the closed convex hull of its extreme points.* Our reformulation is the following: *Every point of a compact convex subset X of a locally convex space is the barycenter of a probability measure on X which is supported by the closure of the extreme points of X.* To prove the equivalence of these two assertions, suppose the former holds and that x is in X. Let Y be the closure of the extreme points of X; then x is in the closed convex hull of Y. By Proposition 1.2, then, x is the barycenter of a probability measure μ on Y. If we extend μ (in the obvious way) to X, we get the desired result. Conversely, suppose the second assertion is valid and that x is in X. Then (defining Y as above) by Proposition 1.2, x is in the closed convex hull of Y, hence in the closed convex hull of the extreme points of X.

It is now clear that any representation theorem using measures supported by the extreme points of X (rather than by their closure) is a sharpening of the Krein-Milman theorem. In fact, Klee [50] has shown that in a sense (which he makes precise) almost every compact convex subset of an infinite dimensional Banach space *is* the closure of its extreme points. For such sets, then, the Krein-Milman representation gives no more information than the "point mass" representation.

The problem of finding measures supported by the extreme points of X arises mainly from the fact that the set of extreme points need not be a Borel set [9, p. 327]. This difficulty is avoided in case X is metrizable, as shown by the following result.

PROPOSITION 1.3 *If X is a metrizable, compact convex subset of a topological vector space, then the extreme points of X form a G_δ set.*

PROOF. Suppose that the topology of X is given by the metric d, and for each integer $n \geq 1$ let $F_n = \{x : x = 2^{-1}(y + z),\ y$ and z in $X,\ d(y, z) \geq n^{-1}\}$. It is easily checked that each F_n is closed, and that a point x of X is not extreme if and only if it is in some F_n. Thus, the complement of the extreme points is an F_σ.

Recall that we always have the trivial representing measure ε_x for a point x of X. If x is not an extreme point of X, then it is

easily seen that there exist other representing measures. Indeed, the
extreme points of X are *characterized* by the fact that they have no
other representing measures.

PROPOSITION 1.4 (BAUER [4]) *Suppose that X is a compact convex subset of a locally convex space E and that $x \in X$. Then x is
an extreme point of X if and only if the point mass ε_x is the only
probability measure on X which represents x.*

PROOF. Suppose that x is an extreme point of X and that the
measure μ represents x. We want to show that μ is supported by
the set $\{x\}$; for this, it suffices (due to the regularity of μ) to show
that $\mu(D) = 0$ for each compact set D with $D \subset X \setminus \{x\}$. Suppose
$\mu(D) > 0$ for some such D; from the compactness of D it follows
that there is some point y of D such that $\mu(U \cap X) > 0$ for every
neighborhood U of y. Choose U to be a closed convex neighborhood
of y such that $K = U \cap X \subset X \setminus \{x\}$. The set K is compact and
convex, and $0 < r = \mu(K) < 1$. (If $\mu(K) = 1$, then the resultant x of
μ would be in K.) Thus, we can define Borel measures μ_1 and μ_2 on
X by $\mu_1(B) = r^{-1}\mu(B \cap K)$ and $\mu_2(B) = (1 - r)^{-1}\mu(B \cap (X \setminus K))$
for each Borel set B in X. Let x_i be the resultant of μ_i; since
$\mu_1(K) = 1$, we see that $x_1 \in K$ and hence $x_1 \neq x$. Furthermore,
$\mu = r\mu_1 + (1 - r)\mu_2$, which implies that $x = rx_1 + (1 - r)x_2$, a
contradiction.

It is interesting to note that Milman's classical "converse" to
the Krein-Milman theorem [28, p. 440] is an easy consequence of
Propositions 1.2 and 1.4; it implies that the closure of ex X is the
smallest closed subset of X which generates X.

PROPOSITION 1.5 *(Milman) Suppose that X is a compact convex
subset of a locally convex space, that $Z \subset X$, and that X is the
closed convex hull of Z. Then the extreme points of X are contained
in the closure of Z.*

PROOF. Indeed, let $Y = \operatorname{cl} Z$ and suppose $x \in \operatorname{ex} X$. By Proposition
1.2, there exists a measure μ on Y which represents x; by Proposition
1.4, $\mu = \varepsilon_x$. It follows that $x \in Y$.

To conclude this introduction, we return to the example of a compact convex subset X of a finite dimensional space E, in order to illustrate the question concerning *uniqueness* of integral representations. Suppose that X is a plane triangle, or more generally, is the convex hull of an affinely independent subset Y of E, that is, X is a simplex. (A set Y is affinely independent provided no point y in Y is in the linear variety generated by $Y \setminus \{y\}$.) It then follows from the affine independence that Y is the set of extreme points of X, and that every element of X has a unique representation by a convex combination of elements of Y. It is not difficult to show (Proposition 10.10) that if X is not a simplex, then some element of X has two such representations. In Section 10 we will give an infinite dimensional generalization of the notion of "simplex" which will allow us to prove (among other things) Choquet's uniqueness theorem, which states that for a metrizable compact convex set X in a locally convex space, each point of X has a *unique* representing measure supported by the extreme points of X if and only if X is a simplex.

In the next section we give an application of the Krein-Milman theorem. Before doing this, it is worthwhile to make some general remarks concerning applications of the various representation theorems. It is generally not difficult to recognize that the objects of interest form a convex subset X of some linear space E. One is then faced with two problems: First, find a locally convex topology for E which makes X compact and at the same time yields sufficiently many continuous linear functionals so that the assertion "μ represents x" has some content. Second, identify the extreme points of X, so that the assertion "μ is supported by the extreme points" has a useful interpretation.

EXERCISE

Prove Carathéodory's sharper form of Minkowski's theorem: If X is a compact convex subset of an n-dimensional space E, then each x in X is a convex combination of at most $n + 1$ extreme points of X. (Hint: Use induction on the dimension. If x is a boundary point of X, there exists a supporting hyperplane H of X with x in $H \cap X$, and the latter set has dimension at most $n - 1$. If x is an interior point of X, choose an extreme point y of X and note that x is in the segment $[y, z]$ for some boundary point z of X.)

2 Application of the Krein-Milman theorem to completely monotonic functions

A real valued function f on $(0,\infty)$ is said to be *completely monotonic* if f has derivatives $f^{(0)} = f, f^{(1)}, f^{(2)}, \ldots$ of all orders and if $(-1)^n f^{(n)} \geq 0$ for $n = 0, 1, 2, \ldots$. Thus, f is nonnegative and nonincreasing, as is each of the functions $(-1)^n f^{(n)}$. [Some examples: $x^{-\alpha}$ and $e^{-\alpha x}$ $(\alpha \geq 0)$.] S. Bernstein proved a fundamental representation theorem for such functions. (See [82] for several proofs and much related material). We will prove the theorem only for *bounded* functions; the extension to unbounded functions (with infinite representing measures) follows from this by classical arguments [82]. We denote the one-point compactification of $[0, \infty)$ by $[0, \infty]$.

THEOREM (Bernstein). *If f is bounded and completely monotonic on $(0, \infty)$, then there exists a unique nonnegative Borel measure μ on $[0, \infty]$ such that $\mu([0, \infty]) = f(0^+)$ and for each $x > 0$,*

$$f(x) = \int_0^\infty e^{-\alpha x} d\mu(\alpha).$$

(Note that the converse is true, since if a function f on $(0, \infty)$ can be represented as above, then differentiation under the integral sign is possible, and it follows that f is completely monotonic. Moreover, by applying the Lebesgue dominated convergence theorem to the functions $\alpha \to e^{-\alpha/n}$ we see that $f(0^+) = \int_0^\infty d\mu = \mu([0, \infty])$, so f is bounded.) The idea of the proof is due to Choquet [16, Ch. VII], who proved this and related results in a much more general setting. We start by giving a sketch of the proof.

Denote by CM the convex cone of all completely monotonic functions f such that $f(0^+) < \infty$. (Since a completely monotonic function f is nonincreasing, this right-hand limit at 0 always exists, although it may be infinite.) Let K be the convex set of those f in CM

such that $f(0^+) \leqq 1$; if $f \in CM$, $f \neq 0$, then $f/f(0^+) \in K$, so it suffices to prove the theorem for elements of K. Now, K is a subset of the space E of all real valued infinitely differentiable functions on $(0, \infty)$, and E is locally convex in the topology of uniform convergence (of functions and all their derivatives) on compact subsets of $(0, \infty)$. We will show that K is compact in this topology, so that the Krein-Milman theorem is applicable to K. Furthermore, the extreme points of K are precisely the functions $x \to e^{-\alpha x}$, $0 \leq \alpha \leq \infty$. [We define $e^{-\infty x}$ to be the zero function on $(0, \infty)$.] It will follow easily that ex K is homeomorphic to $[0, \infty]$ and is therefore compact. By the Krein-Milman theorem, to each f in K there exists a Borel probability measure m on ex K which represents f. The measure m can be carried to a measure μ on $[0, \infty]$ and the evaluation functionals $f \to f(x)$ $(x > 0)$ are continuous on E; these facts are easily combined to obtain the desired representation. The uniqueness assertion is obtained by a simple application of the Stone-Weierstrass theorem to the subalgebra of $C([0, \infty])$ generated by the exponentials.

The first step in our proof is to show that K is a compact subset of E. The topology on E is the same as that given by the countable family of pseudonorms

$$p_{m,n}(f) = \sup\{|f^{(k)}(x)| : m^{-1} \leqq x \leqq m, 0 \leqq k \leqq n\}$$

$(m, n = 1, 2, 3, \dots)$. Thus, E is metrizable, and every closed and bounded subset of E is compact. [This may be proved by Ascoli's theorem, together with repeated use of the diagonal procedure, or by following the outline given in the exercises on "distribution spaces" in [47].] It is easily seen that K is closed. To show that it is bounded we must show that for each m and n, $\sup\{p_{m,n}(f) : f \in K\}$ is finite, and for this it suffices to show that $\sup\{|f^{(n)}(x)| : m^{-1} \leq x \leq m, f \in K\}$ is finite for each $n \geq 0$ and $m \geq 1$. It is clear that the following lemma will establish this fact.

LEMMA 2.1 *Let* $K_n = \{(-1)^n f^{(n)} : f \in K\}$, $n = 0, 1, 2, \dots$. *Then for each $a > 0$ and each $n \geq 0$, the (nonnegative) functions in K_n are bounded above on $[a, \infty)$ by $a^{-n} 2^{(n+1)(n/2)}$.*

PROOF. We proceed by induction. The functions in K_0 are bounded above by 1, so suppose the assertion is true for K_n. Since the functions in K_{n+1} are nonincreasing, it suffices to establish the bound at the point a. By applying the mean value theorem to $f^{(n)}$ on $[a/2, a]$, we see that there exists c with $a/2 < c < a$ such that $(a/2)f^{(n+1)}(c) = f^{(n)}(a) - f^{(n)}(a/2)$. This fact, together with the induction hypothesis (applied at $a/2$), shows that

$$
\begin{aligned}
(a/2)^{-n}2^{(n+1)(n/2)} &\geqq (-1)^n f^{(n)}(a/2) \\
&\geqq (-1)^{n+1}(a/2)f^{(n+1)}(c) \\
&\geqq (a/2)(-1)^{n+1}f^{(n+1)}(a),
\end{aligned}
$$

and the desired result follows.

[A compactness proof different from the above may be obtained by using the topology of *pointwise* convergence on $(0, \infty)$. This is also locally convex, and, of course, the evaluation functionals are continuous. It is known [82, p. 151] that a function is completely monotonic if and only if it satisfies a certain sequence of "iterated difference" inequalities; since these are defined pointwise, it is easily seen that CM is closed in this topology, and the Tychonov product theorem then yields compactness of K.]

Our next step is to identify the extreme points of K.

LEMMA 2.2 *The extreme points of K are those functions f of the form $f(x) = e^{-\alpha x}, x > 0, 0 \leq \alpha \leq \infty$.*

PROOF. Suppose that $f \in \mathrm{ex}\, K$ and that $x_0 > 0$. For $x > 0$, let $u(x) = f(x + x_0) - f(x)f(x_0)$. Suppose that we have shown that $f \pm u \in K$. Since f is extreme, this implies that $u = 0$, so that $f(x + x_0) = f(x)f(x_0)$ whenever $x, x_0 > 0$. Since f is continuous on $(0, \infty)$, this implies that either $f = 0$ (the case $\alpha = \infty$) or $f(x) = e^{-\alpha x}$ for some α. Since $0 \leq -f'(x) = \alpha e^{-\alpha x}$, we must have $\alpha \geq 0$. It remains to show that $f \pm u \in K$. Let $b = f(x_0)$ (so that $0 \leq b \leq 1$), and note that $(f + u)(0^+) = (1 - b)f(0^+) + b \leq 1$ and $(f - u)(0^+) = f(0^+) - b[1 - f(0^+)] \leq f(0^+) \leq 1$. Furthermore,

$$(-1)^n(f + u)^{(n)}(x) = (1 - b)(-1)^n f^{(n)}(x) + (-1)^n f^{(n)}(x + x_0) \geqq 0$$

and $(-1)^n(f - u)^{(n)}(x) =$

$$= [(-1)^n f^{(n)}(x) - (-1)^n f^{(n)}(x + x_0)] + b(-1)^{(n)}f^{(n)}(x).$$

Since $(-1)^n f^{(n)}$ is nonincreasing, the latter is nonnegative.

To prove the reverse inclusion, consider the transformation T_r $(r > 0)$ of K into itself defined by $(T_r f)(x) = f(rx)$. Since T_r is one-to-one, onto, and preserves convex combinations, it carries ex K onto itself. Since K is compact, it is the closed convex hull of its extreme points, and therefore has at least one which is nonconstant. By what we have just proved, this extreme point is of the form $e^{-\alpha x}$ for some $\alpha > 0$, and hence the image $e^{-\alpha r x}$ of this function under T_r is extreme. Since this holds for each $r > 0$, *all* the exponentials are extreme (and the constant functions 0 and 1 are clearly extreme), so the proof is complete.

We now finish the proof of Bernstein's theorem for bounded functions. It is not difficult to show that the map $T: \alpha \to e^{-\alpha(\cdot)}$ from $[0, \infty]$ into K is continuous; since $[0, \infty]$ is compact, its image ex K is also compact. By the Krein-Milman representation theorem, to each f in K there corresponds a regular Borel probability measure m on ex K such that $L(f) = \int_{\text{ex } K} L \, dm$ for each continuous linear functional L on E. Now, if $x > 0$, then the evaluation functional $L_x(f) = f(x)$ is continuous on E, so that $f(x) = \int_{\text{ex } K} L_x \, dm$ for each $x > 0$. Define μ on each Borel subset B of $[0, \infty]$ by $\mu(B) = m(TB)$ (i.e., $\mu = m \circ T$). Since $L_x(T\alpha) = e^{-\alpha x}$, we have

$$
\begin{aligned}
f(x) &= \int_{\text{ex } K} L_x \, dm = \int_{T^{-1}(\text{ex } K)} L_x \circ T \, d(m \circ T) \\
&= \int_0^\infty e^{-\alpha x} d\mu(\alpha) \quad \text{for each} \quad x > 0.
\end{aligned}
$$

It remains to prove that μ is unique. Suppose there exists a second measure ν on $[0, \infty]$ such that $f(x) = \int_0^\infty e^{-\alpha x} d\nu(\alpha)$ $(x > 0)$ and $\nu([0, \infty]) = f(0^+)$. For each $x \geq 0$ the function $\alpha \to e^{-\alpha x}$ is continuous on $[0, \infty]$. Let A be the subalgebra of $C([0, \infty])$ generated by these functions; A consists of finite linear combinations of the same functions and as linear functionals on $C[0, \infty]$, μ and ν are equal on A. Since A separates points of $[0, \infty]$, the Stone-Weierstrass theorem implies that it is dense in $C([0, \infty])$, so $\mu = \nu$.

3 Choquet's theorem: The metrizable case.

In this section we will prove Choquet's representation theorem for metrizable X. This is actually a special case of the general Choquet-Bishop-de Leeuw theorem, but its proof is quite short and it gives us an opportunity to introduce some of the machinery which is needed in the main result.

Suppose that h is a real valued function defined on a convex set C. The function h is *affine* [*convex*] if $h[\lambda x + (1 - \lambda)y] = [\leq] \lambda h(x) + (1 - \lambda)h(y)$ for each x, y in C and $0 \leq \lambda \leq 1$. We say that h is *concave* if $-h$ is convex, and h is called *strictly convex* if h is convex and the defining inequality is strict whenever $x \neq y$ and $0 < \lambda < 1$. Recall that a real-valued function f is said to be *upper semicontinuous* if for each real λ, $\{x: f(x) < \lambda\}$ is open, while it is *lower semicontinuous* if $-f$ is upper semicontinuous.

Denote by A the set of all continuous affine functions on X. Note that A is a subspace of the Banach space $C(X)$ and that A contains the constant functions. Furthermore, A contains all functions of the form $x \to f(x) + r$, where $f \in E^*$, r is real and $x \in X$, so that A contains sufficiently many functions to separate the points of X.

DEFINITION. *If f is a bounded function on X and $x \in X$, let $\bar{f}(x) = \inf\{h(x) : h \in A \text{ and } h \geq f\}$.*

The function $\bar{f}$, which is called the *upper envelope* of f, has the following useful properties:

(a) $\bar{f}$ is concave, bounded, and upper semicontinuous (hence Borel measurable).

(b) $f \leq \bar{f}$ and if f is concave and upper semicontinuous, then $f = \bar{f}$.

(c) If f, g are bounded, then $\overline{f + g} \leq \bar{f} + \bar{g}$ and $|\bar{f} - \bar{g}| \leq \|f - g\|$, while $\overline{f + g} = \bar{f} + g$ if $g \in A$. If $r > 0$, then $\overline{rf} = r\bar{f}$.

The proofs of most of the above facts follow in a straight-forward manner from the definitions. The second assertion in (a) follows from the fact that constant functions are affine. The second assertion in (b) may be proved as follows: If f is concave and upper semicontinuous, then in the locally convex space $E \times R$ the set $K = \{(x, r) : f(x) \geq r\}$ (i.e., the set of points below the graph of f) is closed and convex. If $f(x_1) < \bar{f}(x_1)$ at some point x_1, the separation theorem asserts the existence of a continuous linear functional L on $E \times R$ which strictly separates $(x_1, \bar{f}(x_1))$ from K, i.e., there exists λ such that $\sup L(K) < \lambda < L(x_1, \bar{f}(x_1))$. From the fact that $L(x_1, f(x_1)) < L(x_1, \bar{f}(x_1))$, it follows that $L(0, 1) > 0$, and hence the function h defined on X by $h(x) = r$ if $L(x, r) = \lambda$ exists and is in A. Furthermore, $f < h$ and $h(x_1) < \bar{f}(x_1)$, a contradiction. The second assertion in (c) again uses the fact that constant functions are affine: Since $f \leq \|f\|$, we have $\bar{f} \leq \|f\|$. Furthermore

$$\bar{f} = \overline{(f - g) + g} \leq \overline{(f - g)} + \bar{g},$$

so $\bar{f} - \bar{g} \leq \overline{f - g}$ and hence $\bar{f} - \bar{g} \leq \|f - g\|$. Interchanging f and g yields the desired result.

THEOREM (Choquet). *Suppose that X is a metrizable compact convex subset of a locally convex space E, and that x_0 is an element of X. Then there is a probability measure μ on X which represents x_0 and is supported by the extreme points of X.*

PROOF. (Bonsall [11]). Since X is metrizable, $C(X)$ (and hence A) is separable. Thus, we can choose a sequence of functions $\{h_n\}$ in A such that $\|h_n\| = 1$, and the set $\{h_n\}_{n=1}^{\infty}$ is dense in the unit sphere of A; in particular, it separates points of X. Let $f = \Sigma 2^{-n} h_n^2$; this limit exists uniformly, hence is in $C(X)$ and it is a strictly convex function in $C(X)$. (Indeed, if $x \neq y$, then $h_n(x) \neq h_n(y)$ for some n, so the affine function h_n is nonconstant on the segment $[x, y]$. It follows that h_n^2 is strictly convex on $[x, y]$ and therefore f is strictly convex on $[x, y]$.) Let B denote the subspace $A + Rf$ of $C(X)$ generated by A and f. Now, from property (c) above, it follows that the functional p defined on $C(X)$ by $p(g) = \bar{g}(x_0)$ ($g \in C(X)$) is subadditive and satisfies $p(rg) = rp(g)$ if $r \geq 0$. Define a linear functional on B by $h + rf \rightarrow h(x_0) + r\bar{f}(x_0)$ (h in A, r real). We will show that this functional is dominated on B by the functional p, i.e., that $h(x_0) + r\bar{f}(x_0) \leqq \overline{(h + rf)}(x_0)$ for each h in A, r in R. If $r \geq 0$, then $\overline{h + rf} = \bar{h} + r\bar{f}$, by (b) and (c), while if $r < 0$, then $h + rf$ is concave, and hence $\overline{h + rf} = h + rf \geq \bar{h} + r\bar{f}$. By the Hahn-Banach theorem, then, there exists a linear functional m on $C(X)$ such that $m(g) \leqq \bar{g}(x_0)$ for g in $C(X)$, and $m(h + rf) = h(x_0) + r\bar{f}(x_0)$ if $h \in A$, $r \in R$. If $g \in C(X)$ and $g \leqq 0$, then $0 \geqq \bar{g}(x_0) \geqq m(g)$, i.e., m is nonpositive on nonpositive functions and hence is continuous. By the Riesz representation theorem, there exists a nonnegative regular Borel measure μ on X such that $m(g) = \mu(g)$ for g in $C(X)$. Since $1 \in A$, we see that $1 = m(1) = \mu(1)$, so μ is a probability measure. Furthermore, $\mu(f) = m(f) = \bar{f}(x_0)$. Now, $f \leqq \bar{f}$, so $\mu(f) \leqq \mu(\bar{f})$. On the other hand, if $h \in A$ and $h \geq f$, then $h \geq \bar{f}$, and consequently $h(x_0) = m(h) = \mu(h) \geqq \mu(\bar{f})$. It follows from the definition of $\bar{f}$ that $\bar{f}(x_0) \geqq \mu(\bar{f})$, and therefore $\mu(f) = \mu(\bar{f})$. This last fact implies that μ vanishes on the complement of $\mathcal{E} = \{x : f(x) = \bar{f}(x)\}$. We complete the proof by showing that $\mathcal{E}$ is contained in the set of extreme points of X. Indeed, if $x = \frac{1}{2}y + \frac{1}{2}z$, where y and z are distinct points of X, then the strict convexity of f implies that $f(x) < \frac{1}{2}(y) + \frac{1}{2}(z) \leqq \frac{1}{2}\bar{f}(y) + \frac{1}{2}\bar{f}(z) \leqq \bar{f}(x)$.

It is interesting to note (and will be useful later) that $\{x : \bar{f}(x) = f(x)\}$ actually *coincides* with the set of extreme points of X. This is a consequence of the next proposition.

DEFINITION. *If μ and λ are probability measures such that $\mu(f) = \lambda(f)$ for each f in A, we will write $\mu \sim \lambda$.*

PROPOSITION 3.1 *If f is a continuous function on the compact convex set X, then for each x in $X, \bar{f}(x) = \sup\{\int f \, d\mu : \mu \sim \varepsilon_x\}$. Consequently, $\bar{f}(x) = f(x)$ if x is an extreme point of X.*

PROOF. The second assertion follows from Proposition 1.4. To prove the first assertion, let $f'(x) = \sup\{\mu(f) : \mu \sim \varepsilon_x\}$; we must show that $f' = \bar{f}$. It follows easily from the definition that f' is concave; we prove that it is upper semicontinuous. Indeed, suppose that $\{x_\alpha\}$ is a net in X converging to a point x, with each $f'(x_\alpha) \geq r$, say. To see that $f'(x) \geq r$, suppose that $\varepsilon > 0$ and for each α choose $\mu_\alpha \sim \varepsilon_{x_\alpha}$ such that $\mu_\alpha(f) > r - \varepsilon$. By weak*-compactness, there exist a probability measure μ and a subnet $\{\mu_\beta\}$ of $\{\mu_\alpha\}$ which converges weak* to μ. If g is in A, then $g(x_\beta) = \mu_\beta(g) \to \mu(g)$; since $g(x_\beta) \to g(x)$, we see that $\mu \sim \varepsilon_x$. Thus, $r - \varepsilon \leq \lim \mu_\beta(f) = \mu(f) \leq f'(x)$; it follows that $f'(x) \geq r$. Since f' is upper semicontinuous, $\{(x, r) : f'(x) \geq r\}$ is closed (and convex) in $E \times R$; using the same argument as in (b) (above), we conclude that $\bar{f} \leq f'$. On the other hand, if h is in A, x in X, and $h \geq f$, then for any $\mu \sim \varepsilon_x$, we have $h(x) = \mu(h) \geq \mu(f)$. It follows that $f'(x) \leq h(x)$, and from this we get $f' \leq \bar{f}$.

4   The Choquet-Bishop-de Leeuw existence theorem

Suppose that X is a nonmetrizable compact convex subset of a locally convex space E. As shown by examples in Bishop-de Leeuw [9], the extreme points of X need not form a Borel set. Thus, the statement "the probability measure μ is supported by the extreme points of X" is meaningless under our present definitions. There are at least two ways to get around this. We can drop the requirement that μ be a Borel measure (i.e., allow measures defined on a different σ-ring), or we can change the definition of "supported by" for Borel measures. An alternative definition might require that μ vanish on every Borel set which is disjoint from the set of extreme points, but Bishop and de Leeuw have shown that it is not always possible to obtain representing measures μ with this property. If, however, one demands only that μ vanish on the *Baire* subsets of X which contain no extreme points, then a representation theorem can be obtained. (Recall that the Baire sets are the members of the σ-ring generated by the compact G_δ sets.) Furthermore, this result leads easily to an equivalent theorem in which the definition of "supported by" remains formally the same, but the measure is no longer a Borel measure.

THEOREM (Choquet-Bishop-de Leeuw). *Suppose that X is a compact convex subset of a locally convex space E, and that x_0 is in X. Then there exists a probability measure μ on X which represents x_0 and which vanishes on every Baire subset of X which is disjoint from the set of extreme points of X.*

The rest of this section is devoted mainly to the proof of this theorem.

DEFINITION. *The set of extreme points of X will be denoted by* ex X. *The set of all continuous affine [convex] functions on X will be denoted by A [C].*

The subspace $C - C$ (of all functions of the form $f - g$, f, g in C) is a *lattice* under the usual partial ordering in $C(X)$. [Note that $\max(f_1 - g_1, f_2 - g_2) = \max(f_1 + g_2, f_2 + g_1) - (g_1 + g_2)$.] Since it contains A, $C - C$ separates the points of X and contains the constant functions; by the Stone-Weierstrass theorem, it is dense in the norm topology of $C(X)$. We now partially order the nonnegative measures on X in the following way:

DEFINITION. *If λ and μ are nonnegative regular Borel measures on X, write $\lambda \succ \mu$ if $\lambda(f) \geqq \mu(f)$ for each f in C.*

This relation is clearly transitive and reflexive; the fact that $\lambda \succ \mu$ and $\mu \succ \lambda$ imply $\lambda = \mu$ comes from the fact that $C - C$ is dense in $C(X)$. Note that if f is in A, then both f and $-f$ are in C, so that $\lambda \succ \mu$ implies $\lambda(f) = \mu(f)$, i.e., λ and μ represent the same linear functional on the subspace A. (In particular, if they are probability measures, then they have the same resultant in X.) It is also well worth noting that if $\mu \sim \varepsilon_x$, then $\mu \succ \varepsilon_x$; indeed, if $f \in -C$, then $\bar{f} = f$ and hence

$$f(x) = \inf\{h(x) : h \in A, h \geqq f\} = \inf\{\mu(h) : h \in A, h \geqq f\} \geqq \mu(f).$$

We will be concerned with measures which are *maximal* with respect to this ordering; such a measure will be called a "maximal measure", without further reference to the ordering. The fact that if $\lambda \succ \mu$, then λ has its support "closer" to the extreme points of X than does μ, may be heuristically verified by considering measures and convex functions on a triangle in the plane, say. This fact is what leads us to hope that a maximal measure will be supported by the extreme points.

LEMMA 4.1 *If λ is a nonnegative measure on X, then there exists a maximal measure μ such that $\mu \succ \lambda$.*

PROOF. Suppose $\lambda \geqq 0$ and let $Z = \{\mu : \mu \geqq 0 \text{ and } \mu \succ \lambda\}$. Suppose that we have found an element μ in Z which is maximal (with respect to the ordering $\succ$) in Z. Then μ will be a maximal measure, since if ν is a nonnegative measure and $\nu \succ \mu$, then $\nu \succ \lambda$, so that $\nu \in Z$ and hence $\nu = \mu$. To find a maximal element of Z, let W be a chain in Z. We may regard W as a net (the directed "index set" being the elements of W themselves) which is contained in the weak* compact set $\{\mu \geq 0$ and $\mu(1) = \lambda(1)\}$. Thus, there exists μ_0 with $\mu_0 \geq 0$ and a subnet $\{\mu_\alpha\}$ of W which converges to μ_0 in the weak* topology. If μ_1 is any element in W, it follows from the definition of subnet that eventually $\mu_\alpha \succ \mu_1$ and hence $\mu_0 \succ \mu_1$. Thus, μ_0 is an upper bound for W; furthermore, since $\mu_0 \succ \lambda$, we have $\mu_0 \in Z$. By Zorn's lemma, then, Z contains a maximal element.

Bishop and de Leeuw originated the idea of looking at maximal measures, although they used an ordering which differs slightly from the one used here. The notion is applied in a very simple way: If x_0 is in X, choose a maximal measure μ such that $\mu \succ \varepsilon_{x_0}$. As noted above, μ represents x_0; it remains to show that the maximality of μ implies that μ vanishes on Baire sets which contain no extreme points. The first step toward doing this is contained in the following result.

PROPOSITION 4.2 *If μ is a maximal measure on X, then $\mu(f) = \mu(\bar{f})$ for each continuous function f on X.*

PROOF. Choose f in $C(X)$ and define the linear functional L on the one-dimensional subspace Rf by $L(rf) = r\mu(\bar{f})$. Define the sublinear functional p on $C(X)$ by $p(g) = \mu(\bar{g})$. If $r \geq 0$, then $L(rf) = p(rf)$, while if $r < 0$, then $0 = \overline{rf - rf} \leq \overline{rf} + \overline{(-rf)} = \overline{rf} - r\bar{f}$, and hence $L(rf) = \mu(r\bar{f}) \leq \mu(\overline{rf}) = p(rf)$. Thus, $L \leq p$ on Rf, and therefore (by the Hahn-Banach theorem), there exists an extension L' of L to $C(X)$ such that $L' \leq p$. If $g \leq 0$, then $\bar{g} \leq 0$, so $L'(g) \leq p(g) = \mu(\bar{g}) \leq 0$. It follows that $L' \geq 0$ and hence there exists a nonnegative measure ν on X such that $L'(g) = \nu(g)$ for each g in $C(X)$. If g is convex, then $-g$ is concave and $-g = \overline{-g}$, so $\nu(-g) \leq p(-g) = \mu(\overline{-g}) = \mu(-g)$, i.e., $\mu \prec \nu$. Since μ is maximal, we must have $\mu = \nu$, and therefore $\mu(f) = \nu(f) = L(f) = \mu(\bar{f})$, which completes the proof.

As we will see later (Proposition 10.3) the converse to the above result is true. More importantly, note that the proposition implies the following: *If μ is a maximal measure, then μ is supported by $\{x : \bar{f}(x) = f(x)\}$, for each f in C.* As shown by Proposition 3.1,

each of these sets contains the extreme points of X. If C contained a
strictly convex function f_0, we would have (as in Choquet's theorem)
$\operatorname{ex} X = \{x : \bar{f}_0(x) = f_0(x)\}$, and the proof would be complete.
Hervé [42] has shown, however, that the existence of a strictly convex
continuous function on X implies that X is metrizable. About the
best we can do in the nonmetrizable case is prove that $\operatorname{ex} X$ *is the
intersection of all the sets of the form* $\{x : \bar{f}(x) = f(x)\}$, f *in* C.
Indeed, if $\bar{f}(x) = f(x)$ for each f in C, and if $x = \frac{1}{2}(y + z)$, $y, z \in X$,
then

$$f(y) + f(z) \geqq 2f(x) = 2\bar{f}(x) \geqq \bar{f}(y) + \bar{f}(z) \geqq f(y) + f(z),$$

i.e., $2f(x) = f(y) + f(z)$ for each f in C. It follows that the same
equality holds for any f in $-C$, hence for each element of $C - C$.
Since the latter subspace is dense in $C(X)$, we must have $x = y = z$,
i.e., x is an extreme point of X.

    To show that any maximal measure μ vanishes on the Baire sets
which are disjoint from $\operatorname{ex} X$, it suffices to show that $\mu(D) = 0$ if D is
a compact G_δ set which is disjoint from $\operatorname{ex} X$. (This is a consequence
of regularity: If B is a Baire set and μ is a nonnegative regular Borel
measure, then $\mu(B) = \sup\{\mu(D) : D \subset B, D$ a compact $G_\delta\}$.) It
will be helpful later if we merely assume that D is a compact *subset*
of a G_δ set which is disjoint from $\operatorname{ex} X$. To show that $\mu(D) = 0$, we
first use Urysohn's lemma to choose a nondecreasing sequence $\{f_n\}$
of continuous functions on X with $-1 \leqq f_n \leqq 0$, $f_n(D) = -1$ and
$\lim f_n(x) = 0$ if $x \in \operatorname{ex} X$. We then show that if μ is maximal, then
$\lim \mu(f_n) = 0$; it is immediate from this that $\mu(D) = 0$. To obtain
this "limit" result requires two slightly technical lemmas. The first
of these is quite interesting, since it reduces the desired result to
Choquet's theorem for metrizable X, using an idea due to P. A.
Meyer. (More precisely, we will use the fact that for each x in X,
there exists $\mu \sim \varepsilon_x$ which is supported by $\operatorname{ex} X$. Since it is not
generally true that every f in A can be extended to an element of
E^*, this is formally stronger than the stated version of Choquet's
theorem. See Proposition 4.5.)

LEMMA 4.3 *Suppose that* $\{f_n\}$ *is a bounded sequence of concave upper semicontinuous functions on* X, *with* $\liminf f_n(x) \geqq 0$ *for each*
x *in* $\operatorname{ex} X$. *Then* $\liminf f_n(x) \geqq 0$ *for each* x *in* X.

PROOF. Assume first that X is metrizable. If x is in X, choose a probability measure $\mu \sim \varepsilon_x$ which is supported by ex X. By hypothesis, $\liminf f_n \geq 0$ a.e. μ, so by Fatou's lemma, $\liminf \mu(f_n) \geq 0$. Since each f_n is concave and upper semicontinuous, assertion (b) in Section 3 shows that $f_n = \bar{f}_n$, so that $f_n(x) = \inf\{h(x) : h \in A, h \geq f_n\} = \inf\{\mu(h) : h \in A, h \geq f_n\} \geq \mu(f_n)$. Thus, $\liminf f_n(x) \geq \liminf \mu(f_n) \geq 0$. Turning to the general case, suppose x is in X, and for each n choose h_n in A such that $h_n \geq f_n$ and $h_n(x) < f_n(x) + n^{-1}$. Let R^N be the countable product of lines with the product topology and define $\phi : X \to R^N$ by $\phi(y) = \{h_n(y)\}$. The function ϕ is affine and continuous, so $X' = \phi(X)$ is a compact convex subset of the metrizable space R^N. Let π_n be the usual "n-th coordinate" projection of R^N onto R; if y is in X, then $\pi_n(\phi y) = h_n(y)$. If x' is in X', the set $\phi^{-1}(x')$ is compact and convex in X; by the Krein-Milman theorem it has an extreme point y. Assuming that x' is in ex X', a simple argument shows that y is in ex X. Since $\pi_n(x') = h_n(y) \geq f_n(y)$, we have $\liminf \pi_n(x') \geq \liminf f_n(y) \geq 0$, for each x' in ex X'. The functions π_n are affine and continuous on the metrizable set X', so from the first part of this proof we conclude that $\liminf \pi_n(x') \geq 0$ for each x' in X'. Taking $x' = \phi(x)$, we obtain $0 \leq \liminf \pi_n(\phi x) = \liminf h_n(x) = \liminf f_n(x)$, which completes the proof.

LEMMA 4.4 *If μ is a maximal measure on X, and if $\{f_n\}$ is a nondecreasing sequence in $C(X)$ such that $-1 \leq f_n \leq 0$ $(n = 1, 2, \ldots)$ and $\lim f_n(x) = 0$ for each x in ex X, then $\lim \mu(f_n) = 0$.*

PROOF. Consider the sequence $\{\bar{f}_n\}$ of concave upper semicontinuous functions. Since $-1 \leq f_n \leq \bar{f}_n \leq 0$, we have $\lim \bar{f}_n(x) = 0$ if x is in ex X; in addition, the sequence $\{\bar{f}_n\}$ is also nondecreasing (and bounded above by zero), so that $\lim \bar{f}_n(x)$ exists for each x in X. It follows from Lemma 4.3 that $\lim \bar{f}_n(x) = 0$ for each x in X. From the Lebesgue bounded convergence theorem it follows that $\lim \mu(\bar{f}_n) = 0$; from Proposition 4.2 we have $\mu(\bar{f}_n) = \mu(f_n)$, which completes the proof.

Thus, we have shown that any maximal measure on X vanishes on the Baire subsets of $X \setminus$ ex X. We have also shown something slightly different: *A maximal measure μ vanishes on any G_δ subset of X contained in $X \setminus \text{ex}\, X$.* (Indeed, we showed that $\mu(D) = 0$ if D is any compact subset of such a set.) This is important, since it shows, in particular, that a maximal measure is supported by any closed set which contains $\text{ex}\, X$, and hence the Choquet-Bishop-de Leeuw theorem generalizes the Krein-Milman theorem.

We next formulate the Choquet-Bishop-de Leeuw theorem in a manner which can perhaps be more convenient for applications.

THEOREM (Bishop-de Leeuw). *Suppose that X is a compact convex subset of a locally convex space, and denote by S the σ-ring of subsets of X which is generated by $\text{ex}\, X$ and the Baire sets. Then for each point x_0 in X there exists a nonnegative measure μ on S with $\mu(X) = 1$ such that μ represents x_0 and $\mu(\text{ex}\, X) = 1$.*

PROOF. By the Choquet-Bishop-de Leeuw theorem there exists a Borel measure λ which represents x_0 and which vanishes on the Baire subsets of $X \setminus \text{ex}\, X$. We need only extend λ to a nonnegative measure μ on S and show that $\mu(\text{ex}\, X) = 1$. To do this, observe that any set S in S is of the form $[B_1 \cap \text{ex}\, X] \cup [B_2 \cap (X \setminus \text{ex}\, X)]$, where B_1 and B_2 are Baire sets. If we let $\mu(S) = \lambda(B_1)$, then μ is well defined and $\mu(\text{ex}\, X) = \lambda(X) = 1$.

As we remarked earlier, not every function in A is of the form $x \to f(x) + r$, f in E^*, r in R. Consider the following example:

Let E be the Hilbert space ℓ^2 in its weak topology, let X be the set of sequences $x = \{x_n\}$ such that $|x_n| \leq 2^{-n}$ and define f on X by $f(x) = \Sigma x_n$. Then f is in A and $f(0) = 0$, but there is no point y in ℓ^2 such that $f(x) = (x, y)$ for all x in X.

This example shows that the subspace $M = E^*|_X + R$ of $C(X)$ may be a proper subspace of A. Nevertheless, the two notions "$\mu \sim \varepsilon_x$" and "μ represents x" (for a probability measure μ on X and a point x in X) coincide, as the following proposition implies.

PROPOSITION 4.5 *The subspace M (defined above) of affine functions is uniformly dense in the closed subspace A of all affine continuous functions on X.*

PROOF. It is evident that the space A is uniformly closed. Suppose that $g \in A$ and $\varepsilon > 0$, and consider the following two subsets of

$E \times R$: $J_1 = \{(x, r) : x \in X, r = g(x)\}$ and $J_2 = \{(x, r) : x \in X$ and $r = g(x) + \varepsilon\}$. These sets are compact, convex, nonempty, and disjoint. By a slightly extended version of the usual separation theorem (obtained by separating the origin from the closed convex difference set $J_2 - J_1$) there exist a continuous linear functional L on $E \times R$ and λ in R such that $\sup L(J_1) < \lambda < \inf L(J_2)$. If we define f on E by the equation $L(x, f(x)) = \lambda$, it follows that f is affine and continuous, and that $g(x) < f(x) < g(x) + \varepsilon$ for x in X, which completes the proof.

5 Applications to Rainwater's and Haydon's theorems

In this section we present two nontrivial results which are nice applications of the Choquet-Bishop-deLeeuw theorem.

To introduce the first one, suppose that Y is a compact Hausdorff space and that f, f_n $(n = 1, 2, 3, \ldots)$ are functions in $C(Y)$. A classical theorem states that $\{f_n\}$ converges weakly to f if and only if the sequence $\{f_n\}$ is uniformly bounded and $\lim f_n(y) = f(y)$ for each y in Y. If we recall that the extreme points of the unit ball U of $C(Y)^*$ are the functionals of the form $f \to \pm f(y)$, then this result is seen to be a special case of the following theorem.

THEOREM (Rainwater [66]). *Let E be a normed linear space and suppose that x, x_n $(n = 1, 2, 3, \ldots)$ are elements of E. Then the sequence $\{x_n\}$ converges weakly to x if and only if $\{x_n\}$ is bounded and $\lim f(x_n) = f(x)$ for each extreme point f of the unit ball U of E^*.*

PROOF. Let Q denote the natural isometry of E into E^{**}. If $\{x_n\}$ converges weakly to x, then for each f in E^*, the sequence of real numbers $(Qx_n)(f)$ is bounded and hence the uniform boundedness theorem shows that $\{Qx_n\}$, hence $\{x_n\}$, is bounded in norm. To prove the converse, suppose that $\{Qx_n\}$ is bounded and that $f(x_n) = (Qx_n)(f) \to (Qx)(f) = f(x)$ for each f in ex U, and that g is an arbitrary element of U. It suffices to show that $(Qx_n)(g) \to (Qx)(g)$. Now, in the weak* topology on E^*, U is compact (and convex) so by the Bishop-de Leeuw theorem there exists a σ-ring $\mathcal{S}$ of subsets of U (with ex $U \in \mathcal{S}$) and a probability measure μ on $\mathcal{S}$ such that $\mu(U \setminus$ ex $U) = 0$ and such that $L(g) = \int L \, d\mu$ for each weak* continuous affine function L on U. In particular, $(Qx_n)(g) = \int Qx_n \, d\mu$ and $(Qx)(g) = \int Qx \, d\mu$. Furthermore, $\{Qx_n\}$ converges to Qx on U a.e. μ, so by the Lebesgue bounded convergence theorem $\int Qx_n \, d\mu \to \int Qx \, d\mu$, and the proof is complete.

Our second application deals with arbitrary Banach spaces.

THEOREM (Haydon [41]). *Let E be a real Banach space and let K be a weak* compact convex subset of E^* such that ex K is norm separable. Then K is the norm closed convex hull of its extreme points (and hence is itself norm separable).*

PROOF. Let $M = \sup\{\|f\| : f \in K\}$, suppose that $\epsilon > 0$ and let $\{f_i\}$ be a norm dense subset of ex K. For each i, let B_i denote the intersection with K of the closed ball of radius $\epsilon/3$ centered at f_i. Thus, each B_i is weak* compact and convex and $\bigcup B_i \supset$ ex K. Let f be a point of K and let μ be a maximal probability measure on K with resultant $r(\mu) = f$. Since $\bigcup B_i$ is a weak* F_σ-set, we have $\mu(\bigcup B_i) = 1$. Let n be a positive integer such that, if $D = \bigcup_{i=1}^n B_i$, then $\mu(D) > 1 - \frac{\epsilon}{3M}$. Then μ can be decomposed as $\mu = \lambda \mu_1 + (1 - \lambda)\mu_2$, where $\lambda = \mu(D)$ and μ_1, μ_2 are probability measures on K defined by

$$\lambda \mu_1 = \mu|_D \quad \text{and} \quad (1 - \lambda)\mu_2 = \mu|_{(K \setminus D)}.$$

(If $\lambda = 1$, let μ_2 be an arbitrary probability measure on K.) Then $f = r(\mu) = \lambda r(\mu_1) + (1 - \lambda)r(\mu_2)$. Since $r(\mu_2) \in K$ we have

$$\|f - \lambda r(\mu_1)\| = (1 - \lambda)\|r(\mu_2)\| \le \frac{\epsilon}{3M} \cdot M = \frac{\epsilon}{3}.$$

Since μ_1 is a probability measure supported by $\bigcup_{i=1}^n B_i$, the point $r(\mu_1)$ lies in the convex hull of $\bigcup_{i=1}^n B_i$, which is weak* compact. Hence $r(\mu_1) = \sum_{i=1}^n \lambda_i g_i$, where $g_i \in B_i$, $\lambda_i \ge 0$ and $\sum_{i=1}^n \lambda_i = 1$. Let $h = \sum_{i=1}^n \lambda_i f_i$. This is a point of co(ex K) and $\|r(\mu_1) - h\| \le \epsilon/3$. Consequently,

$$\|f - h\| \le \|f - \lambda r(\mu_1)\| + (1 - \lambda)\|r(\mu_1)\| + \|r(\mu_1) - h\| \le \frac{\epsilon}{3} + \frac{\epsilon}{3} + \frac{\epsilon}{3}.$$

Thus, co(ex K) is norm dense in K.

6 A new setting: The Choquet boundary

In the Introduction, the Riesz representation theorem was reformulated as a representation theorem of the Choquet type. Although the *conclusion* of the Riesz theorem is quite sharp (for each element of the convex set X under consideration there exists a unique representing measure supported by $\operatorname{ex} X$), the *hypotheses* restrict its application to a very special class of compact convex sets. In what follows we will (among other things) describe a related family of sets which appears to be only slightly larger than that involved in the Riesz theorem, but which actually "contains" all the sets which interest us, in the sense that every compact convex subset of a locally convex space is affinely homeomorphic to a member of the family.

Throughout this section, Y will denote a compact Hausdorff space, and $C_c(Y)$ will denote the space of all continuous complex-valued functions on Y, with supremum norm. (We continue to denote the space of *real*-valued continuous functions on Y by $C(Y)$.) In order to work with the complex Banach space $C_c(Y)$, we recall some basic facts about *any* complex Banach space E. As usual, the dual of the locally convex space E^* in its weak* topology is E itself, with each $x \in E$ defining a weak* continuous linear functional $f \to f(x)$ on E^*. In looking at convex sets in $(E^*, weak^*)$, one considers it as a real vector space whose dual consists of all functionals of the form $f \to \operatorname{Re} f(x)$, $x \in E$.

DEFINITION. *Suppose that M is a linear subspace (not necessarily closed) of $C(Y)$ (or of $C_c(Y)$) and that $1 \in M$. The state space $K(M)$ of M is the set of all L in M^* such that $L(1) = 1 = \|L\|$.*

If M^* is taken in its weak* topology, then $K(M)$ is a nonempty compact convex subset of a locally convex space, and the results from preceding sections are applicable. Note that the Riesz theorem dealt with the set $K(C(Y))$. In order to make full use of these results, it is necessary to have descriptions of the extreme points of $K(M)$. There is not a great deal to be said in the general case, but (as will be shown later) for the case when M is a subalgebra of $C_c(Y)$, Bishop and Bishop-de Leeuw have obtained useful and interesting characterizations of ex $K(M)$.

Since we will consider the real and complex cases simultaneously, it will be of help to recall one form of the Riesz representation theorem for $C_c(Y)^*$: If L is in $C_c(Y)^*$, then there exist nonnegative regular Borel measures μ_1, μ_2, μ_3 and μ_4 on Y such that the measure $\mu = \mu_1 - \mu_2 + i(\mu_3 - \mu_4)$ represents L and the total variation $\|\mu\|$ of μ equals $\|L\|$. If $L(1) = 1 = \|L\|$, then $L \geq 0$ (that is, $Lf \geq 0$ whenever $f \geq 0$) and hence $\mu = \mu_1 \geq 0$. [A simple proof of the above assertion about L goes as follows: If $f \geq 0$, let D be any closed disc in the complex plane which contains the bounded set $f(Y)$; assume D has center α and radius $r > 0$. Then $\|f - \alpha\| \leq r$ so $r \geq |L(f - \alpha)| = |L(f) - \alpha|$, i.e., $Lf \in D$. Thus, Lf is in the closed convex hull of $f(Y)$ (which is the intersection of all such discs) and since $f(Y)$ is contained in the nonnegative real axis, we have $Lf \geq 0$.] It follows that even in the complex case, the functionals in $K(M)$ may be represented by probability measures on Y; indeed, if $L \in K(M)$, then it may be extended to an element of $K(C_c(Y))$ by the (complex form of the) Hahn-Banach theorem. By the Riesz theorem there exists a complex measure μ on Y such that $L(f) = \int_Y f \, d\mu$ for each f in M. It follows from the previous remarks that μ is a probability measure.

DEFINITION. *If y is in Y, let ϕy be the element of $K(M)$ defined by $(\phi y)(f) = f(y)$, f in M. Note that ϕ is continuous from Y into the weak* topology on $K(M)$.*

If M *separates points* of Y, then ϕ is one-to-one, and hence is a homeomorphism, embedding Y as a compact subset of $K(M)$. If $L \in K(M)$ and μ is a measure on Y such that $L(f) = \mu(f)$ for f in M, then we can "carry" μ to a measure μ' on $K(M)$ in the obvious way: $\mu' = \mu \circ \phi^{-1}$. Since M is the conjugate space to M^* (in its weak* topology), it follows easily that μ' represents L.

LEMMA 6.1 *Suppose that M is a subspace of $C(Y)$ (or of $C_c(Y)$) and that $1 \in M$. Then $K(M)$ equals the weak* closed convex hull of*

$\phi(Y)$.

PROOF. If the above assertion is false, then there exists f in M such that $\sup\{(\mathrm{Re}\, f)(y) : y \in Y\} < \sup\{\mathrm{Re}\, L(f) : L \in K(M)\} \leq \sup\{\mathrm{Re}\, L(f) : L \in C(Y)^*, \|L\| = 1\} = \|\mathrm{Re}\, f\|$. We may assume (by adding a positive constant to f) that $\mathrm{Re}\, f \geq 0$; the first term then becomes $\|\mathrm{Re}\, f\|$ and we have a contradiction.

DEFINITION. *Suppose that M is a linear subspace of $C(Y)$ (or of $C_c(Y)$) and that $1 \in M$. Let $B(M)$ be the set of all y in Y for which ϕy is an extreme point of $K(M)$. We call $B(M)$ the* Choquet boundary *for M.*

The reason for introducing this notion is apparent: An element L in $K(M)$ is an extreme point of $K(M)$ if and only if $L = \phi y$ for some y in $B(M)$. [The "if" part of this assertion comes from the definition of $B(M)$; on the other hand, Lemma 6.1 and Milman's theorem (Section 1) imply that $\mathrm{ex}\, K(M) \subset \phi(Y)$.] We have the following "intrinsic" characterization of $B(M)$ in terms of measures on Y, at least for subspaces M which separate the points of Y.

PROPOSITION 6.2 *Suppose that M is a subspace of $C(Y)$ (or of $C_c(Y)$) which separates the points of Y and contains the constant functions. Then y is in the Choquet boundary $B(M)$ of M if and only if $\mu = \varepsilon_y$ is the only probability measure on Y such that $f(y) = \int_Y f\, d\mu$ for each f in M.*

PROOF. Suppose that $y \in B(M)$ and suppose that for some measure μ on Y, $f(y) = \int f\, d\mu$ for each f in M. Then the measure $\mu' = \mu \circ \phi^{-1}$ is defined on (the Borel subsets of) $K(M)$, and the above relation means that $\mu' \sim \varepsilon_{\phi y} = \varepsilon_y \circ \phi^{-1}$. Since $\phi y \in \mathrm{ex}\, K(M)$, Proposition 1.4 implies $\mu \circ \phi^{-1} = \varepsilon_y \circ \phi^{-1}$, and since ϕ is a homeomorphism we have $\mu = \varepsilon_y$. Conversely, suppose $y \notin B(M)$, so that $\phi y \notin \mathrm{ex}\, K(M)$. Then there exist distinct functionals in $K(M)$, and hence distinct measures μ_1 and μ_2 on Y which represent them, such that $(\phi y)(f) = \frac{1}{2}\mu_1(f) + \frac{1}{2}\mu_2(f)$ for each f in M. Let $\mu = \frac{1}{2}\mu_1 + \frac{1}{2}\mu_2$; then $\mu(f) = f(y)$ for each f in M, although $\mu \neq \varepsilon_y$. (Indeed, suppose $\varepsilon_y = \mu = \frac{1}{2}\mu_1 + \frac{1}{2}\mu_2$. Since μ_1 and μ_2 are distinct, $\mu_1 \neq \varepsilon_y$ and hence $\mu_1\{y\} < 1$. It follows that $\mu\{y\} < 1$, i.e., $\mu \neq \varepsilon_y$.)

(The above characterization is actually the *definition* of the Choquet boundary used by some authors. If one wishes to use this definition for subspaces M which do not separate points, a more complicated version is necessary [9, 30].) This proposition makes it evident that when M is all of $C(Y)$ or $C_c(Y)$, we have $\phi(Y) = \operatorname{ex} K(M)$; equivalently, $B(M) = Y$. An example where $B(M) \neq Y$ may be constructed as follows: Let $Y = [0,1]$ and let $M = \{f \in C(Y) : f(\frac{1}{2}) = \frac{1}{2}f(0) + \frac{1}{2}f(1)\}$. Then $B(M) = Y \setminus \{\frac{1}{2}\}$. [Clearly, $\frac{1}{2} \notin B(M)$. If $x \neq \frac{1}{2}$ and $\mu(f) = f(x)$ for each f in M, then by choosing functions in M which "peak" at x, it can be seen that $\mu = \varepsilon_x$.]

DEFINITION. *Suppose that M is a subspace of $C(Y)$ or of $C_c(Y)$ and suppose $1 \in M$. A subset B of Y is said to be a boundary for M if for each f in M there exists a point x in B such that $|f(x)| = \|f\| \ (= \sup\{|f(y)| : y \in Y\})$. If there is a smallest closed boundary for M (i.e., a closed boundary which is contained in every closed boundary), it is called the Šilov boundary for M.* (For some illuminating examples, see the end of Section 8.)

PROPOSITION 6.3 *Suppose that M is a subspace of $C(Y)$ (or $C_c(Y)$) with $1 \in M$. If $f \in M$, then there exists y in the Choquet boundary $B(M)$ such that $|f(y)| = \|f\|$, i.e., $B(M)$ is a boundary for M.*

PROOF. Let L_0 be any element of $K(M)$ such that $|L_0(f)| = \|f\|$ (for instance, evaluation at some point where $|f|$ attains its maximum) and let K_0 be the set of all L in $K(M)$ such that $L(f) = L_0(f)$. The set K_0 is nonempty, weak* compact, and convex, hence it has an extreme point L_1 which, since K_0 is itself an extremal subset (or *face*) of $K(M)$, is necessarily an extreme point of $K(M)$. Hence $L_1 = \phi y$ for some y in $B(M)$, and $|f(y)| = |L_1(f)| = |L_0(f)| = \|f\|$.

PROPOSITION 6.4 *Suppose that M is a subspace of $C(Y)$ (or $C_c(Y)$) which contains the constant functions and separates points of Y. Then the closure of the Choquet boundary is the Šilov boundary for M.*

PROOF. It follows from Proposition 6.3 that the closure of $B(M)$ is a closed boundary for M. It remains to show that if B is a closed boundary for M, then $B(M) \subset B$ (and hence $\text{cl} \, B(M) \subset B$). Suppose not; then there exists y in $B(M) \setminus B$ and hence a neighborhood U of y with $U \subset Y \setminus B$. We will show that there exists f in M such that $\sup |f(Y \setminus U)| < \sup |f(U)|$; this will imply that B is not a boundary for M, a contradiction. The remainder of the proof is due (for the real case) to Choquet. Note that ϕy is an element of $\text{ex} \, K(M)$ and that $\phi(U)$ is a (relative) weak* neighborhood of ϕy in $\phi(Y)$. Using the definition of the weak* topology and the fact that $1 \in M$, we can find $f_1, \ldots, f_n$ in M and $\varepsilon > 0$ such that $\phi y \in \cap \{L : \text{Re} \, L(f_i) < \varepsilon\} \cap \phi(Y) \subset \phi(U)$. [This isn't immediately obvious. We can certainly find a finite number of functions g_j in M such that

$$\phi(y) \in \cap\{L : |\text{Re} \, L(g_j) - \text{Re} \, g_j(y)| < \varepsilon\} \cap \phi(Y) \subset \phi(U).$$

If we replace each g_j by $f_j \equiv g_j - \text{Re} \, g_j(y)$, the first intersection above is

$$\cap\{L : |\text{Re} \, f_j| < \varepsilon\} = \cap\{L : \text{Re} \, L(f_j) < \varepsilon\} \cap \{L : -\text{Re} \, L(f_j) < \varepsilon\},$$

so we have simply doubled the number of functions we started with.] Consider the compact convex sets $K_i = \{L : \text{Re} \, L(f_i) \geq \varepsilon\} \cap K(M)$, $i = 1, \ldots, n$. The convex hull J of their union is again a compact convex subset of $K(M)$, but does *not* contain the extreme point ϕy; otherwise, ϕy would be a convex combination of elements L_i of K_i, $i = 1, \ldots, n$. Since $\phi y \notin J$, we can apply the separation theorem to obtain a function f in M such that $\sup \text{Re} \, f(J) < \text{Re} \, (\phi y)(f)$. Since $\phi(Y) \setminus \phi(U) \subset \cup K_i \subset J$, we have $\sup(\text{Re} \, f)(Y \setminus U) < \text{Re} \, f(y)$. By adding a sufficiently large positive constant to f we get the desired result.

We next show that every nonempty compact convex subset of a locally convex space is of the form $K(M)$ (for suitable Y and M).

PROPOSITION 6.5 *If X is a compact convex subset of a locally convex space E, then there exists a separating subspace M of $C(X)$, with $1 \in M$, such that X is affinely homeomorphic with $K(M)$.*

PROOF. Let M be those functions in $C(X)$ of the form $g(x) = f(x) + r$, where f is in E^*, r in R. Define ϕ from X to $K(M)$ as before; it is

easily checked that ϕ is affine. To see that $\phi(X) = K(M)$, suppose
that L is in $K(M)$; by using the Hahn-Banach and Riesz theorems
as above, we can find a measure μ on X such that $L(g) = \mu(g)$ for
each g in M. By Proposition 1.1, μ has a unique resultant x in X;
it follows that $\phi x = L$.

It follows from the foregoing discussion that we can carry problems concerning representing measures into the context of function
spaces and Choquet boundaries. This latter setting has been aptly
referred to as "the Bishop-de Leeuw setup." One advantage of the
Bishop-de Leeuw setup is the relative ease with which examples may
be constructed. [Another advantage is that it lends itself to the discussion of function algebras, which was Bishop's [8] original motivation for proving a special case of the Choquet theorem; see Section
8.]

We conclude this section with a form of the representation theorem which is due to Bishop and de Leeuw. In order to do this (and
for later purposes) we observe that for separating subspaces M there
is a suitable definition of $\lambda \succ \mu$ for measures λ and μ on Y, namely,
define $\lambda \succ \mu$ to mean that $\lambda \circ \phi^{-1} \succ \mu \circ \phi^{-1}$. If we are given a
measure μ on Y and we want a maximal measure λ with $\lambda \succ \mu$,
we can choose a maximal measure λ' on $K(M)$ with $\lambda' \succ \mu \circ \phi^{-1}$.
In view of the remarks in Section 4 (prior to the Bishop-de Leeuw
theorem), λ' is supported by the compact set $\phi(Y)$, hence is of the
form $\lambda \circ \phi^{-1}$ for a (maximal) measure λ on Y such that $\lambda \succ \mu$. To
see that λ vanishes on the Baire subsets of $Y \setminus B(M)$, we need only
show that it vanishes on any compact G_δ subset $D \subset Y \setminus B(M)$.
Now, $\phi(D)$ is a G_δ in $\phi(Y)$ and it misses $\mathrm{ex}K(M)$, hence the same is
true of $A = \phi(D) \cup [K(M) \setminus \phi(Y)]$. It follows that the complement
of A is an F_σ in $\phi(Y)$ and therefore is an F_σ in $K(M)$, so A is a G_δ
in $K(M)$ which misses $\mathrm{ex}K(M)$. By the remarks following Lemma
4.4, λ' vanishes on $A \supset \phi(D)$, hence $\lambda(D) = \lambda'(\phi(D)) = 0$.

THEOREM *Suppose that M is a subspace of $C(Y)$ (or of $C_c(Y)$)
which separates points and contains the constant functions. If $L \in
M^*$, then there exists a complex measure μ on Y such that $L(f) =
\int_Y f \, d\mu$ for each f in M and $\mu(S) = 0$ for any Baire set S in Y
which is disjoint from the Choquet boundary for M.*

PROOF. By applying the Hahn-Banach and Riesz theorems, we may obtain a measure $\lambda = \lambda_1 - \lambda_2 + i(\lambda_3 - \lambda_4)$ on Y such that $L(f) = \lambda(f)$ for each f in M. For each i we can find a maximal measure μ_i on $K(M)$ with $\mu_i \succ \lambda_i$. We know that μ_i vanishes on the Baire sets which are disjoint from $B(M)$, and $\mu_i(f) = \lambda_i(f)$ for f in M. If we define $\mu = \mu_1 - \mu_2 + i(\mu_3 - \mu_4)$, we get a measure with the required properties.

7 Applications of the Choquet boundary to resolvents

Let X be a compact Hausdorff space, and suppose that for each $\lambda > 0$ there is a linear transformation $R_\lambda : C(X) \to C(X)$ such that $R_\lambda \geq 0$ (i.e., $R_\lambda f \geq 0$ whenever $f \geq 0$) and $R_\lambda 1 = 1/\lambda$. We call the family of operators $R_\lambda (\lambda > 0)$ a *resolvent* if the following identity is valid for all λ, $\lambda' > 0$:

$$(*) \qquad\qquad R_{\lambda'} - R_\lambda = (\lambda - \lambda') R_{\lambda'} R_\lambda.$$

[Such families arise in the study of Markov processes. If $(T_t : t > 0)$ is a semigroup of Markov operators from $C(X)$ into itself (i.e., $T_s T_t = T_{s+t}$, $T_t 1 = 1$, $T_t \geq 0$), then under suitable conditions

$$(R_\lambda f)(x) = \int_0^\infty e^{-\lambda t}(T_t f)(x)dt \qquad (x \text{ in } X, \lambda > 0)$$

exists for all f in $C(X)$ and defines a resolvent. Under certain hypotheses, every resolvent is obtainable in this way from a semigroup of Markov operators, and the content of this section is a convergence theorem related to the proof of this result. (See the papers [55] and [68] for more detailed information on this subject.) None of the facts in this paragraph are needed to follow the exposition given below, which is due to Lion [55] and was originally shown us by Choquet.]

We first prove some elementary facts which follow easily from the definition of a resolvent.

1. *For each $\lambda > 0$, R_λ is continuous and $\|R_\lambda\| = 1/\lambda$. Indeed, if $f \in C(X)$, then $\pm f \leq \|f\| \cdot 1$, hence $\pm R_\lambda f \leq \|f\| \cdot R_\lambda 1 = (1/\lambda)\|f\|$, so $\|R_\lambda\| \leq 1/\lambda$. But $R_\lambda 1 = 1/\lambda$.*

2. *For each λ and λ', $R_\lambda R_{\lambda'} = R_{\lambda'} R_\lambda$. This is trivial if $\lambda = \lambda'$ and follows from $(*)$ otherwise.*

3. *The operators R_λ have the same range.* Given λ, let $M_\lambda = R_\lambda[C(X)]$ be the range of R_λ. For any λ and λ', if $f \in C(X)$, then $R_\lambda f - R_{\lambda'} f = (\lambda' - \lambda) R_\lambda (R_{\lambda'} f)$, so $R_{\lambda'} f \in M_\lambda$. Thus, $M_{\lambda'} \subset M_\lambda$.

Let M denote the common range of the operators R. *Throughout the remainder of this section, we assume that M separates points of X.* (Even if we did not assume this, it would still be possible to formulate a suitable theorem analogous to the one below, but the statement would be unnecessarily complicated.) The next theorem is essentially the same as one originally proved by Ray [68], using a different method.

THEOREM *Suppose that X is a compact Hausdorff space, that R_λ is a resolvent on $C(X)$, and that M is the common range of the operators R_λ.*

1. *If x is in the Choquet boundary B of M, then for all f in $C(X)$,*

$$\lim_{\lambda \to \infty} \lambda(R_\lambda f)(x) = f(x).$$

2. *If x is in X, there exists a regular Borel measure μ_x on X such that, for each f in $C(X)$,*

$$\lim_{\lambda \to \infty} \lambda(R_\lambda f)(x) = \int_X f \, d\mu_x.$$

The measure μ_x is supported by B, in the sense that $\mu_x(A) = 0$ for any Baire set $A \subset X \setminus B$.

3. *If x is in X and the conclusion to (1) holds, then $x \in B$.*

PROOF. We first show that if g is in M, then $\|\lambda R_\lambda g - g\| \to 0$ as $\lambda \to \infty$. Indeed, we can write $g = R_1 f$ for some f in $C(X)$, and hence $\lambda R_\lambda g - g = \lambda R_\lambda R_1 f - R_1 f = 1 \cdot R_\lambda R_1 f - R_\lambda f = R_\lambda(R_1 f - f)$. It follows that

$$\|\lambda R_\lambda g - g\| \leqq \|R_\lambda\| \, \|R_1 f - f\| \leqq (1/\lambda)\|R_1 f - f\| \to 0.$$

Suppose, now, that $x \in X$ and $\lambda > 0$. The functional defined on $C(X)$ by $f \to \lambda(R_\lambda f)(x)$ is nonnegative on nonnegative functions and takes 1 into 1, hence there exists a probability measure $\mu_{x,\lambda}$ on X such that $\lambda(R_\lambda f)(x) = \int f \, d\mu_{x,\lambda}$ for each f in $C(X)$. For each $\lambda_0 > 0$, let $A(\lambda_0, x)$ be the closure (in the weak* topology of $C(X)^*$) of $\{\mu_{x,\lambda} : \lambda \geq \lambda_0\}$. For fixed x, the sets $A(\lambda_0, x)(\lambda_0 > 0)$ form a nested family of nonempty compact sets and hence have nonempty intersection $A(x)$. We next show that if $\mu \in A(x)$, then $\mu \sim \varepsilon_x$, i.e., $\mu(g) = g(x)$ for each g in M. Indeed, given $\varepsilon > 0$, there exists $\lambda_0 > 0$ such that $|\lambda(R_\lambda g)(x) - g(x)| \leq \|\lambda R_\lambda g - g\| < \varepsilon/2$ for $\lambda \geq \lambda_0$. Since $\mu \in A(\lambda_0, x)$, every weak* neighborhood of μ contains a measure $\mu_{x,\lambda}$ for some $\lambda \geq \lambda_0$. In particular, then, $|\mu_{x,\lambda}(g) - \mu(g)| < \varepsilon/2$ for some $\lambda \geq \lambda_0$. It follows that $|\mu(g) - g(x)| < \varepsilon$ for all $\varepsilon > 0$.

Suppose now that $x \in B$, the Choquet boundary for M. The previous remark shows that if $\mu \in A(x)$, then $\mu \sim \varepsilon_x$, and from the uniqueness property of the Choquet boundary we conclude that if $x \in B$, then $A(x) = \{\varepsilon_x\}$. Hence if $x \in B$ and U is any weak* neighborhood of ε_x, we must have $A(\lambda_0, x) \subset U$ for some λ_0, so that $\mu_{x,\lambda} \in U$ if $\lambda \geq \lambda_0$. Thus, $\mu_{x,\lambda}$ converges to ε_x as $\lambda \to \infty$, i.e., for each x in B, $\lim \lambda(R_\lambda f)(x)$ exists and equals $f(x)$ for each f in $C(X)$, which proves (1).

Suppose, next, that $x \in X$. By the Choquet-Bishop-de Leeuw theorem there exists a maximal measure $\mu_x \sim \varepsilon_x$ on X. Given f in $C(X)$ and $\lambda > 0$, let $g_\lambda = \lambda R_\lambda f$; then $g_\lambda \in M$ so that $g_\lambda(x) = \mu_x(g) = \int_X g \, d\mu_x$. Now $\|g_\lambda\| = \|\lambda R_\lambda f\| \leq \lambda \|R_\lambda\| \cdot \|f\| = \|f\|$ and, for y in B, $g_\lambda(y) = \lambda(R_\lambda f)(y) \to f(y)$, by what was just proved. Suppose, now, that X is metrizable. Then $\mu_x(B) = 1$ so $g_\lambda \to f$ a.e. μ_x, and the Lebesgue dominated convergence theorem implies that $\lim \lambda(R_\lambda f)(x) = \mu_x(f)$. If X is not metrizable, we can (as in the proof of the Bishop-de Leeuw theorem) extend μ_x to the σ-ring generated by B and the Baire subsets of X, so that $\mu_x \sim \varepsilon_x$ and $\mu_x(B) = 1$. All the functions involved are measurable with respect to this larger σ-ring and we can apply the dominated convergence theorem as before.

It remains to show that $x \in B$ whenever $\lim \lambda(R_\lambda f)(x) = f(x)$ for all f in $C(X)$. Suppose, then, that $\mu \sim \varepsilon_x$ (and hence $\mu \succ \varepsilon_x$); we must show that $\mu = \varepsilon_x$. The above proof that $\lim \lambda(R_\lambda f)(x) = \mu_x(f)$ is valid for any maximal measure μ_x such that $\mu_x \sim \varepsilon_x$. Thus, we can suppose that μ_x is a maximal measure satisfying $\mu_x \succ \mu$. Since $\mu_x(f) = \lim \lambda(R_\lambda f)(x) = f(x)$ for all f in $C(X)$, we have $\varepsilon_x = \mu_x \succ \mu \succ \varepsilon_x$, and the proof is complete.

8   The Choquet boundary for uniform algebras

By a *uniform algebra* (or *function algebra*) in $C_c(Y)$ (Y compact
Hausdorff) we mean any uniformly closed subalgebra of $C_c(Y)$ which
contains the constant functions and separates points of Y. For
metrizable Y, the Choquet boundary of a uniform algebra A has
a particularly simple description (Bishop [8]): It consists of the *peak
points* for A, i.e., of those y in Y for which there exists a function f
in A with the property that $|f(x)| < |f(y)|$ if $x \neq y$. This result is
a special case of a characterization (for arbitrary Y) due to Bishop
and de Leeuw [9], which is the main theorem of this section.

DEFINITION. *Suppose that A is a uniform algebra in $C_c(Y)$ and that
$y \in Y$. We say that y satisfies:*

   Condition I—*if for any open neighborhood U of y and any $\varepsilon > 0$
there exists f in A such that $\|f\| \leq 1$, $|f(y)| > 1 - \varepsilon$ and $|f| \leq \varepsilon$ in
$Y \setminus U$;*

   Condition II—*if, whenever S is a G_δ containing y, there exists
f in A such that $|f(y)| = \|f\|$ and $\{x : |f(x)| = \|f\|\} \subset S$.*

THEOREM (Bishop-de Leeuw). *Suppose that A is a uniform algebra
in $C_c(Y)$ and that $y \in Y$. The following assertions are equivalent:*

(i) *The point y satisfies Condition I.*

(ii) *For each open set U containing y, there exists $f \in A$ such that
$|f(y)| = \|f\|$ and $|f| < \|f\|$ in $Y \setminus U$.*

(iii) *For each $x \in Y$ with $y \neq x$, there exists $f \in A$ such that
$|f(x)| < |f(y)| = \|f\|$.*

(iv) *The point y satisfies Condition II*

(v) *The point y is in the Choquet boundary $B(A)$ of A.*

PROOF. (i) implies (ii). Suppose that y satisfies Condition I and that U is an open set containing y. We will construct a sequence $\{g_n\}$ of functions in A with the following properties:

(a) $\|g_{n+1} - g_n\| \leq 2^{-n+1}$

(b) $\|g_n\| \leq 3(1 - 2^{-n-1})$

(c) $g_n(y) = 3(1 - 2^{-n})$

(d) $|g_{n+1} - g_n| < 2^{-n-1}$ in $Y \backslash U$.

Suppose, for the moment, that we have done this. By (a), the sequence $\{g_n\}$ converges to a function $f \in A$; (b) and (c) imply that $\|f\| = 3 = f(y)$. Moreover, if $x \in Y \backslash U$, then, writing $f = g_n + \sum_{k=n}^{\infty}(g_{k+1} - g_k)$, we have

$$|f(x)| \leq \|g_n\| + \sum_{k=n}^{\infty} |g_{k+1}(x) - g_k(x)| <$$

$$< 3(1 - 2^{-n-1}) + \sum_{k=n}^{\infty} 2^{-k-1} < 3.$$

We obtain the sequence $\{g_n\}$ by induction: Since $y \in U$, Condition I implies that there exists f in A such that $\|f\| \leq 1$, $|f(y)| > \frac{3}{4}$ and $|f| \leq \frac{1}{4}$ in $Y \backslash U$. Let $g_1 = \left(\frac{3}{2}\right)[f(y)]^{-1}f$. Since $|f(y)| > \frac{3}{4}$, $|g_1| \leq \frac{3}{2} \cdot \frac{4}{3} = 2 < 3(1 - 2^{-2})$, so g_1 satisfies (b). Also, $|g_1(y)| = \frac{3}{2} = 3(1 - 2^{-1})$, so g_1 satisfies all the relevant conditions. Suppose that $g_1, g_2, \ldots, g_k$ have been chosen to satisfy the above four conditions. Since $g_k(y) = 3(1 - 2^{-k})$, there is a neighborhood V of y, $V \subset U$, such that $|g_k| < 3(1 - 2^{-k}) + 2^{-k-2}$ in V. We can choose another function f in A such that $\|f\| \leq 1$, $|f(y)| > \frac{3}{4}$ and $|f| \leq \frac{1}{4}$ in $Y \backslash V$. Define $h = (3 \cdot 2^{-k-1})[f(y)]^{-1}f$; then $h(y) = 3 \cdot 2^{-k-1}$ and $\|h\| \leq 2^{-k+1}$. Also, for x in $Y \backslash V$, $|h(x)| < 2^{-k-1}$. Let $g_{k+1} = g_k + h$; properties (a), (c) and (d) are immediate. To check (b), suppose $x \in V$; then $|g_{k+1}(x)| \leq |g_k(x)| + |h(x)| \leq 3(1-2^{-k}) + 2^{-k-2} + 2^{-k+1} = 3(1 - 2^{-k-2})$. On the other hand, if $x \in Y \backslash V$, then $|g_{k+1}(x)| \leq \|g_k\| + 2^{-k-1} \leq 3(1 - 2^{-k-1}) + 2^{-k-1} = 3 - 2^{-k} < 3(1 - 2^{-k-2})$. This completes the induction and the proof that (i) implies (ii)..

That (ii) implies (iii) is immediate. To see that (iii) implies (iv), that is, that y satisfies Condition II, suppose that S is a G_δ set containing y. Let $\{U_n\}$ be a decreasing sequence of open sets with $S = \cap U_n$. For each n we will find f_n in A such that $\|f_n\| = 1 = f_n(y)$ and $|f_n| < 1$ in $Y \setminus U_n$. Once this is done, the function $f = \Sigma 2^{-n} f_n$ will satisfy the properties of Condition II. Suppose, then, that $n \geq 1$ and that $x \in Y \setminus U_n$. By (iii), there exists f_x in A such that $f_x(y) = 1 = \|f_x\|$ and $|f_x| < 1$ in a neighborhood V_x of x. By compactness of $Y \setminus U_n$ we can choose a finite number $f_{x_1}, \ldots, f_{x_k}$ of such functions for which $V_{x_1}, \ldots, V_{x_k}$ cover $Y \setminus U_n$. The function $f_n = k^{-1} \Sigma f_{x_i}$ then has the required properties.

To prove that (iv) implies (v), suppose that y satisfies Condition II. We will show that $\mu = \varepsilon_y$ is the only probability measure on Y such that $\mu(f) = f(y)$ for each f in A; from Proposition 6.2, we can conclude that $y \in B(A)$. Indeed, suppose that μ is such a measure; to see that $\mu(\{y\}) = 1$, it suffices to show that $\mu(S) = 1$ for any G_δ set which contains y. If S is such a set, choose f in A such that $y \in \{x : |f(x)| = \|f\|\} \subset S$; then $\|f\| = |f(y)| = |\mu(f)| \leq \int_S |f| d\mu + \int_{Y \setminus S} |f| d\mu \leq \|f\| \mu(S) + \int_{Y \setminus S} |f| d\mu$. If $\mu(Y \setminus S) > 0$, then $\int_{Y \setminus S} |f| d\mu < \|f\| \mu(Y \setminus S)$, a contradiction which completes the proof that (iv) implies (v).

To show that (v) implies (i), we need a simple lemma.

LEMMA 8.1 *If M is a separating subspace of $C_c(Y)$, with $1 \in M$, then* Re M *(the space of real parts of functions in M) is a separating subspace of $C(Y)$, and $B(\mathrm{Re}\, M) = B(M)$.*

PROOF. Use Proposition 6.2 and the fact that for a real measure μ on Y, $\mu(\mathrm{Re}\, f) = (\mathrm{Re}\, f)(y)$ for every f in M if and only if $\mu(f) = f(y)$ for every f in M.

We return to the proof that (v) implies (i): Suppose $y \in B(A) = B(\mathrm{Re}\, A)$, that U is an open neighborhood of y, and that $0 < \varepsilon < 1$. Choose a function g in $C(Y)$ such that $0 \leq g \leq 1$, $g(y) = 1$ and $g = 0$ in $Y \setminus U$. Denote the weak* compact convex set $K(\mathrm{Re}\, A) \subset (\mathrm{Re}\, A)^*$ by X and use Tietze's extension theorem to obtain f in $C(X)$ such that $f = g \circ \phi^{-1}$ on $\underline{\phi(Y)} \subset X$. Since $\phi y \in \mathrm{ex}\, X$, it follows from Proposition 3.1 that $\overline{(-f)}(\phi y) = (-f)(\phi y) = -g(y) = -1$. Now the space of continuous affine functions on X is (by Proposition 4.5) isomorphic to the uniform closure of the functions in $\operatorname{Re} A$, and therefore $\overline{(-f)}(\phi y) = \inf\{h(y) : h \in \operatorname{Re} A, h \geq -f\}$. It follows that there exists h_0 in $\operatorname{Re} A$ such that $h_0 \leq g$ and $h_0(y) > \log(\delta - 1)/\log \delta$ (where $\delta = 1/\varepsilon$). Let $h = (\log \delta)(h_0 - 1)$; then $h \in \operatorname{Re} A$ and there exists k in $\operatorname{Re} A$ such that $h + ik \in A$. Since A is closed in $C_c(Y)$, the function $f_1 = \exp(h + ik)$ is in A. Since $|f_1| = e^h$, it follows easily that $|f_1| \leq 1$, $|f_1(y)| > 1 - \varepsilon$ and $|f_1| \leq \varepsilon$ in $Y \setminus U$.

Note that if Y is metrizable, then each point of Y is a G_δ, and the equivalence between (iv) and (v) yields the following corollary.

COROLLARY 8.2 (Bishop [8]). *If Y is metrizable and A is a function algebra in $C_c(Y)$, then the Choquet boundary for A coincides with the set of peak points for A.*

The Choquet boundary can be a proper subset of the Šilov boundary, as shown by the following example: Let Y be the unit circle $\{z : |z| = 1\}$ in the complex plane, and let A_1 be those functions in $C_c(Y)$ which are restrictions of functions f which are analytic in $|z| < 1$, continuous in $|z| \leq 1$ and which satisfy $f(0) = f(1)$. It follows from the maximum modulus principle for analytic functions that every point of Y except 1 is a peak point for A_1; since Y is metrizable, this shows that $B(A_1) = Y \setminus \{1\}$, while the Šilov boundary for A_1 is Y.

We give a related example which is the motivation for the term "boundary:" Let $Y = \{z : |z| \leq 1\}$ and let A_2 be the set of all functions in $C_c(Y)$ which are analytic in $|z| < 1$. Then the Choquet boundary coincides with the Šilov boundary and these equal the boundary $\{z : |z| = 1\}$ of Y.

Finally, let $Y = \{z : |z| = 1\}$ and let A_3 be the restrictions to Y of the functions in A_2; then the Choquet and Šilov boundaries equal Y (so this can happen for proper subalgebras of $C_c(Y)$.)

For a description of Bishop's application of the foregoing material to an approximation problem for functions in the complex plane, see Section 16.

It is not generally true that the peak points and the Choquet boundary coincide (in the metrizable case) for linear subspaces M of $C_c(Y)$ which are not algebras. (One inclusion *does* hold: The proofs of "(iii) implies (iv)" and "(iv) implies (v)" in the Bishop-deLeeuw Theorem (above) didn't use the fact that A is an algebra: For any separating linear subspace, every peak point is in the Choquet boundary.) Now consider the following example: Let Y be the subset of the plane consisting of the convex hull of two disjoint circles, and let M be the complex valued affine functions on Y. The four tangent points to the circles are in the Choquet boundary, but are not peak points—a fact most easily seen by considering Re M. In this example, the peak points for M are dense in the Choquet boundary for M, a fact which is true in general and is a corollary to the following classical result concerning Banach spaces. By a *smooth point* of the unit sphere of a Banach space E, we mean a point x, $\|x\| = 1$, for which there is a unique f in E^* such that $\|f\| = 1 = f(x)$.

PROPOSITION 8.3 (S. MAZUR) *Let E be a separable real (or complex) Banach space and let $S = \{x : \|x\| = 1\}$ denote the unit sphere of E. Then the smooth points of S form a dense G_δ subset of S.*

PROOF. In the case of a complex space, we will consider it as a real space (in the usual way); the set of smooth points (which we denote by sm S) is unchanged. We will show that sm S is a countable intersection of dense open subsets of S; since S is a complete metric space, the Baire category theorem will yield the desired conclusion. Let $\{x_n\}$ be a dense sequence in S. For positive integers m and n, let D_{mn} be those x in S such that $f(x_n) - g(x_n) < m^{-1}$ whenever f, g in E^* satisfy $\|f\| = f(x) = 1 = g(x) = \|g\|$. It is easily verified that if $x \in S \setminus \text{sm } S$, then $x \notin D_{mn}$ for some m and n, hence sm $S = \cap D_{mn}$. To see that $S \setminus D_{mn}$ is closed in S, suppose that $y_k \in S \setminus D_{mn}$ and $y_k \to y$. Choose functions f_k, g_k of norm one such that $f_k(y_k) = 1 = g_k(y_k)$ and $f_k(x_n) - g_k(x_n) \geq m^{-1}$, $k = 1, 2, 3, \ldots$. It follows easily from the weak* compactness of the unit ball of E^* that $y \in S \setminus D_{mn}$. It remains to show that each set D_{mn} is dense in S. Suppose not; then for some m and n w can choose y in S and $\delta > 0$ such that $\|x - y\| < \delta$ and $\|x\| = 1$ imply $x \notin D_{mn}$. Let $y_1 = y$ and choose f_1, g_1 in E^* such that $f_1(y_1) = \|f_1\| = 1 = \|g_1\| = g_1(y_1)$ and $f_1(x_n) \geq m^{-1} + g_1(x_n)$. We will proceed by induction to define a sequence $\{y_k\}$ in S and corresponding functionals f_k, g_k of norm one such that $\|y_1 - y_k\| < (1 - 2^{-k})\delta$, $f_k(y_k) = 1 = g_k(y_k)$ and $f_k(x_n) \geq k\, m^{-1} + g_1(x_n)$. Since $f_k(x_n) \leq 1$, this will lead to a contradiction. Suppose we have chosen y_k which has the above properties. We define $y_{k+1} = (y_k + \alpha x_n)/\|y_k + \alpha x_n\|$, where $\alpha > 0$ is chosen to be small enough to insure that $\|y_k - y_{k+1}\| < 2^{-k-1}\delta$. Thus, $\|y_{k+1}\| = 1$ and $\|y_1 - y_{k+1}\| < (1 - 2^{-k})\delta + \|y_k - y_{k+1}\| < (1 - 2^{-k-1})\delta < \delta$. It follows that $y_{k+1} \notin D_{mn}$, so there exist f_{k+1}, g_{k+1} of norm one such that $f_{k+1}(y_{k+1}) = 1 = g_{k+1}(y_{k+1})$ and $f_{k+1}(x_n) \geq m^{-1} + g_{k+1}(x_n)$. Now,

$$1 = \|y_{k+1}\| \geq f_k(y_{k+1}) = [1 + \alpha f_k(x_n)]/\|y_k + \alpha x_n\|.$$

Since $g_{k+1}(y_{k+1}) = 1 \geq g_{k+1}(y_k)$, we have

$$\|y_k + \alpha x_n\| = g_{k+1}(y_k + \alpha x_n) \leq 1 + \alpha g_{k+1}(x_n).$$

These facts combine to show that $f_k(x_n) \leq g_{k+1}(x_n)$, so that

$$f_{k+1}(x_n) \geq m^{-1} + g_{k+1}(x_n) \geq m^{-1} + f_k(x_n) \geq (k+1)m^{-1} + g_1(x_n),$$

and the proof is complete.

Mazur's theorem was the first in an extensive and continuing series of investigations into those properties of Banach spaces E which guarantee that any continuous real-valued convex function on E is differentiable on a dense G_δ subset. See, for instance, [62] and references therein.

COROLLARY 8.4 *Suppose that Y is a compact metrizable space and that M is a uniformly closed separating subspace of $C_c(Y)$ (or of $C(Y)$) which contains the constant functions. Then the peak points for M are dense in the Choquet boundary for M.*

PROOF. Let P be those points y in Y such that $f(y) = \|f\|$ for some smooth point f of the unit sphere of M. It is immediate that every point of P is a peak point for M, and P will be dense in $B(M)$ if $\phi(P)$ is weak* dense in ex $K(M)$. This latter will be true (by Milman's theorem in Section 1) if $K(M)$ is the weak* closed convex hull of $\phi(P)$. If it were not, we could choose g in M, $\|g\| = 1$, such that $\sup \operatorname{Re} g(P) < \sup \operatorname{Re} g(K(M))$. Since the smooth points are uniformly dense in the unit sphere of M, there would exist a smooth point f satisfying this same inequality. For such a function f, however, the left side is $\|f\|$ and the right side is at most $\|f\|$, a contradiction.

Many of the classical theorems in approximation theory can be formulated in terms of convergence of a sequence of linear operators to the identity operator. To illustrate this, we consider one example in detail, obtaining Bernstein's proof of the Weierstrass approximation theorem. For each $n \geq 1$, define the operator B_n from $C[0,1]$ into the polynomials of degree at most n by setting, for $f \in C[0,1]$,

$$(B_n f)(x) = \sum_{k=0}^{n} \binom{n}{k} f(k/n) x^k (1-x)^{n-k}, \quad x \in [0,1].$$

Bernstein proved that $\{B_n f\}$ converges uniformly to f, which obviously gives a constructive proof of the Weierstrass theorem. P. Korovkin [52] observed that each B_n is a positive operator (if $f \geq 0$, then $B_n f \geq 0$) and he proved the following remarkable result.

THEOREM (Korovkin[52]). *Suppose that $\{T_n\}$ is a sequence of positive operators from $C[0,1]$ into itself with the property that $\{T_n f\}$ converges uniformly to f for the three functions $f(x) = x^k$, $k = 0, 1, 2$. Then $\{T_n f\}$ converges uniformly to f for every $f \in C[0,1]$.*

To show that the Bernstein operators satisfy the hypotheses of Korovkin's theorem, we must show that $\{B_n I^k\}$ converges uniformly to I^k for $k = 0, 1, 2$, where $I(x) = x$ for $x \in [0,1]$. Consider the binomial expansion

(1) $$(x + a)^n = \sum_{k=0}^{n} \binom{n}{k} x^k a^{n-k}.$$

Setting $a = 1 - x$ shows that $B_n 1 = 1$ for each n. Differentiating (1) with respect to x twice, multiplying by $\frac{x^2}{n^2}$, setting $a = 1 - x$ and utilizing the previous identities yields

$$B_n I^2 = I^2 + \frac{1}{n}(I - I^2)$$

for each n, so $\{B_n I^2\}$ converges uniformly on $[0,1]$ to I^2.

We won't prove Korovkin's theorem itself, since we are interested in a more general result due to Šaškin [73]. Suppose that X is a compact Hausdorff space and that M is a subset of $C(X)$. We call M a *Korovkin set* in $C(X)$ provided Korovkin's theorem holds for M, that is, provided it is true that $\{T_n f\}$ converges uniformly to f for each $f \in C(X)$ whenever $\{T_n\}$ is a sequence of positive operators on $C(X)$ such that $\{T_n f\}$ converges uniformly to f for each $f \in M$. (Korovin's theorem asserts that $\{1, x, x^2\}$ is a Korovkin set in $C[0,1]$.) Note that M is a Korovkin set if and only if the same is true of its linear span, so we may assume that M is a linear subspace.

THEOREM (Šaškin[73]). *Suppose that X is a compact metrizable space and that M is a linear subspace of $C(X)$ which contains 1 and separates points of X. Then M is a Korovkin set in $C(X)$ if and only if the Choquet boundary $B(M)$ for M is all of X.*

[To see that this does indeed yield Korovkin's theorem we need only observe that for any $x_0 \in [0,1]$, the polynomial $1 - (x - x_0)^2$ peaks at x_0, so the latter point is in the Choquet boundary of the span of $1, x$ and x^2.]

Suppose, first, that $B(M) = X$ and that $\{T_n\}$ is a sequence of positive operators on $C(X)$ such that $\|T_n g - g\| \to 0$ for all $g \in M$. Given $f \in C(X)$, we must show that $\|T_n f - f\| \to 0$; equivalently, we must show that every subsequence of $\{\|T_n f - f\|\}$ has itself a subsequence which converges to 0. For simplicity of notation, assume that $\{\|T_n f - f\|\}$ is the initial subsequence and choose, for each n, a point $x_n \in X$ such that

$$\|T_n f - f\| = |(T_n f)(x_n) - f(x_n)|.$$

By taking a further subsequence we can assume that $x_n \to x$ for some $x \in X$. Define a sequence $\{L_n\}$ of positive linear functionals on $C(X)$ by

$$L_n h = (T_n h)(x_n), \quad h \in C(X).$$

Since $1 \in M$ we have $L_n 1 \to 1$, so we may assume that $L_n 1 > 0$ for each n. Thus, $\frac{L_n}{L_n 1}$ can be considered to be a probability measure

μ_n on X and the sequence $\{\mu_n\}$ necessarily has a subsequence $\{\mu_{n_k}\}$ which converges weak* to a probability measure μ. Now if $g \in M$, then $\|T_{n_k}g - g\| \to 0$ by hypothesis, so from the continuity of g and the inequalities

$$|(T_{n_k}g)(x_{n_k}) - g(x)| \le |(T_{n_k}g)(x_{n_k}) - g(x_{n_k})| + |g(x_{n_k}) - g(x)| \le$$

$$\le \|T_{n_k} - g\| + |g(x_{n_k}) - g(x)|,$$

we now conclude that $\mu_{n_k}(g) \to g(x)$ for $g \in M$. Since we also have $\mu_{n_k}(g) \to \mu(g)$ for all g, we have $\mu(g) = g(x)$ for $g \in M$, that is $\mu \sim \varepsilon_x$. Since $x \in B(M)$ by hypothesis, this implies that $\mu = \varepsilon_x$, so that $\{L_{n_k}/L_{n_k}1\}$ converges weak* to ε_x. This, together with the continuity of f and the inequality

$$\|T_{n_k}f - f\| = |(T_{n_k}f)(x_{n_k}) - f(x_{n_k})| \le$$

$$\le |(T_{n_k}f)(x_{n_k}) - f(x)| + |f(x) - f(x_{n_k})|$$

yields the desired result.

To prove the converse, suppose that M is a Korovkin set and that $x \in X$; we must show that if μ is a probability measure on X with $\mu \sim \varepsilon_x$, then $\mu = \varepsilon_x$. Choose a decreasing sequence $\{U_n\}$ of open neighborhoods of x with $\cap U_n = \{x\}$ and for each n choose $g_n \in C(X)$ such that $0 \le g_n \le 1$, $g_n(x) = 1$ and $g_n(X \setminus U_n) = \{0\}$. For each n define T_n by

$$T_n f = \mu(f)g_n + (1 - g_n)f, \quad f \in C(X).$$

Clearly each T_n is a positive linear operator such that $T_n 1 = 1$. If $g \in M$ and $\epsilon > 0$, we can choose N such that $|g(x) - g(y)| < \epsilon$ for $y \in U_N$. Moreover, $T_n g - g = [g(x) - g]g_n$, so for $n \ge N$ and any $y \in X$ we have

$$|(T_n g)(y) - g(y)| = |g(x) - g(y)|g_n(y) < \epsilon,$$

which shows that $\|T_n g - g\| \to 0$. This being true for every $g \in M$, the hypothesis that M is a Korovkin set implies that $\|T_n f - f\| \to 0$ for every $f \in C(X)$; in particular, $(T_n f)(x) \to f(x)$, so that $\mu(f) = f(x)$ for each $f \in C(X)$, that is, $\mu = \varepsilon_x$, which was to be shown.

Much more about Korovkin–type theorems may be found in Donner [26].

10 Uniqueness of representing measures.

The question of uniqueness of representing measures is a natural one, both in applications and in the theory itself. As always, one must specify clearly the context within which uniqueness is being asserted. What we would like most is a theorem which characterizes those compact convex X with the property that to each point there exists a unique measure that represents the point and is supported by the extreme points of X. Choquet has proved such a theorem for metrizable X, but there is no satisfactory result in the general case. On the other hand, Choquet and Meyer have characterized those X with the property that to each point there corresponds a unique *maximal* measure which represents the point. Since maximal measures are "supported" by the extreme points, it would seem that this answers the question, but the fact that "supported" is taken in an approximate sense makes a considerable difference. An example by Mokobodzki will show that uniqueness of maximal representing measures does *not* imply uniqueness of representing measures which vanish on Baire subsets of $X \setminus \mathrm{ex}\, X$.

   In this section, then, we return to the study of a compact convex set X in a *real* locally convex space E. As before, we denote by $A[C]$ the affine [convex] continuous functions on X. For our present purposes, it will be convenient to assume that X is contained in a closed hyperplane which misses the origin—*this will be assumed throughout this section.* [There is no generality lost in making this assumption, since we may embed E as the hyperplane $E \times \{1\}$ in $E \times R$ (product topology); the image $X \times \{1\}$ of X is affinely homeomorphic with X.] The main reason for doing this is that the question of uniqueness is most naturally studied when X is the *base* of a convex cone P, i.e., when there is a convex cone P (with vertex at the origin) such that $y \in P$ if and only if there exists a unique $\alpha \geq 0$ and x in X such that $y = \alpha x$. If X is contained in a hyperplane which misses the origin, then this is certainly the case; take $P = \tilde{X}$, where $\tilde{X} = \{\alpha x : \alpha \geq 0, x \in X\}$ is the cone generated by X. On the other hand, if X is a base for a cone P, then $0 \notin X$ (otherwise 0 would not have a unique representation in the form αx, $\alpha \geq 0$, $x \in X$), so by the compactness of X and the separation theorem there exists a continuous linear functional f on E and $\beta > 0$ such that $f(x) \geq \beta$ for all $x \in X$. The set $X' = \{x' \in P: \ f(x') = \beta\}$ is also a base for P which is affinely homeomorphic to X under the map $X \ni x \to \beta x / f(x)$. Thus, *whenever a compact convex set is a base for a cone P, we can always assume that it is of the form $H \cap P$ for some closed hyperplane H in E which misses the origin.*

Now, recall that a cone P in E induces a translation invariant partial ordering on E: $x \geq y$ if and only if $x - y \in P$. If P has a base X, then $P \cap (-P) = \{0\}$, so that $x \geq y$ and $y \geq x$ imply $x = y$. Furthermore, if x and y are in the subspace $P - P$ generated by P, then there exists z in P such that $z \geq x$ and $z \geq y$, i.e., x and y have an *upper bound* in $P - P$. We say that z is the *least* upper bound for x and y if $z \leq w$, whenever $w \geq x$ and $w \geq y$, and we denote this least upper bound by $x \vee y$. It is easy to check that the translation invariance of P guarantees that for x, y, z in $P - P$ we have $(x \vee y) + z = (x + z) \vee (y + z)$. If a convex set X (not necessarily compact) is the base of a cone $\tilde{X}$, we call X a *simplex* if the space $\tilde{X} - \tilde{X}$ is a *vector lattice* in the ordering induced by X (that is, if each pair x, y in $\tilde{X} - \tilde{X}$ has a least upper bound $x \vee y$ in $\tilde{X} - \tilde{X}$). Equivalently, $\tilde{X} - \tilde{X}$ is a lattice if and only if each pair x, y has a *greatest lower bound* (definition obvious), which is denoted by $x \wedge y$; we have $x \wedge y = -(-x \vee -y)$. Finally, note that $\tilde{X} - \tilde{X}$ *is a vector lattice if (and only if) $\tilde{X}$ is a lattice.* [Indeed, suppose that each pair $x, y \in \tilde{X}$ has a least upper bound $x \vee y$ in $\tilde{X}$. If $x = x_1 - x_2$ and $y = y_1 - y_2$ are elements of $\tilde{X} - \tilde{X}$, let $z = (x_1 + y_2) \vee (y_1 + x_2) - (x_2 + y_2)$; we will show that z is the least upper bound of x, y in $\tilde{X} - \tilde{X}$. Since $z - x = (x_1 + y_2) \vee (y_1 + x_2) - (x_1 + y_2)$ we have $z \geq x$; similarly, $z \geq y$. If $w = w_1 - w_2 \in \tilde{X} - \tilde{X}$ and $w \geq x$, $w \geq y$, we must show that $w \geq z$. The first two inequalities imply that $w_1 + x_2 + y_2 \geq w_2 + x_1 + y_2$ and $w_1 + x_2 + y_2 \geq w_2 + x_2 + y_1$. Using the translation invariance mentioned earlier, $w - z = (w_1 + x_2 + y_2) - [(w_2 + x_2 + y_1) \vee (w_2 + x_1 + y_2)] \geq 0.$]

It is easy to verify that being a simplex is an "intrinsic" property of X, that is, if X is contained in a hyperplane which misses the origin in E, if X_1 is similarly situated in E_1, and if there exists a one-to-one affine map of X onto X_1, then this map may be extended in the obvious way to a one-to-one, additive, order-preserving map which carries $\tilde{X}$ onto $\tilde{X}_1$, so that one of these cones is lattice if and only if the other is a lattice. At the end of this section we will show that the above definition of a simplex coincides with the usual one in case X is finite dimensional.

The main result of this section is the theorem that each point of X is represented by a *unique* maximal measure if and only if X is a simplex. This result, together with a number of equivalent formulations, is due to G. Choquet and P. Meyer [22], and our proofs follow theirs. Choquet's original uniqueness theorem for metrizable X is an easy corollary.

Let us formulate the uniqueness portion of the Riesz representation theorem in terms of simplices. Suppose that Y is a compact Hausdorff space and let X be the compact convex set of all probability measures on Y. As we noted in the Introduction, the Riesz theorem can be formulated as follows: To each point of X there exists a unique representing measure which is supported by $\operatorname{ex} X = \phi(Y)$. The uniqueness assertion can be considered to be a consequence of the fact that X is a simplex, i.e., that *the cone of all nonnegative measures on Y has the set X of probability measures as a base and is a lattice in the usual ordering.* [We will not prove this well-known fact, but we recall one way of defining the greatest lower bound $\lambda \wedge \mu$ of two nonnegative measures λ and μ. Let $\nu = \lambda + \mu$; then both λ and μ are absolutely continuous with respect to ν, hence have Radon-Nikodym derivatives f and g, respectively. Let $h = \min(f, g)$ (this is defined a.e. ν), and let $\lambda \wedge \mu = h\nu$.]

We need one important technical result concerning lattices, the "Decomposition Lemma," which is assertion (iii) in the following lemma.

LEMMA 10.1 *Suppose that V is a vector lattice.*

(i) $(x + z) \wedge (y + z) = (x \wedge y) + z$ for each x, y, z in V.

(ii) If $x \geqq 0$, $y \geqq 0$ and $z \geqq 0$, then

$$(x + y) \wedge z \leqq (x \wedge z) + (y \wedge z).$$

(iii) If $\{x_i : i \in I\}$ and $\{y_j : j \in J\}$ are finite sequences of nonnegative elements of V, and if $\Sigma_{i \in I} x_i = \Sigma_{j \in J} y_j$, then there exist $z_{ij} \geq 0$, $(i,j) \in I \times J$, such that $x_i = \Sigma_{j \in J} z_{ij}$ $(i \in I)$ and $y_j = \Sigma_{i \in I} z_{ij}$ $(j \in J)$.

PROOF. The fact that the partial ordering in V is translation invariant yields an immediate proof of (i). To see (ii), let $u = (x+y) \wedge z$, so that $u \leq x + y$ and $u \leq z$. Since $0 \leq x$, we have $u \leq x + z$, and hence $u \leq (x+y) \wedge (x+z) = x + (y \wedge z)$. On the other hand, $y \wedge z \geq 0$, so $u \leq z + (y \wedge z)$, and therefore $u \leq [x + (y \wedge z)] \wedge [z + (y \wedge z)] = (x \wedge z) + (y \wedge z)$. To prove (iii) (the Decomposition Lemma), it is not difficult to use induction on the number of elements in I and in J in order to reduce the proof to the case $I = J = \{1, 2\}$. To prove this case, suppose $x_1 + x_2 = y_1 + y_2$ (all elements nonnegative) and let $z_{11} = x_1 \wedge y_1$, $z_{12} = x_1 - z_{11}$, and $z_{21} = y_1 - z_{11}$. These z_{ij} are nonnegative, and $z_{12} \wedge z_{21} = (x_1 - z_{11}) \wedge (y_1 - z_{11}) = (x_1 \wedge y_1) - z_{11} = 0$ (by (i)). Furthermore, $z_{12} + x_2 = x_1 + x_2 - z_{11} = y_1 + y_2 - z_{11} = z_{21} + y_2$; it remains to show that $z_{22} = x_2 - z_{21} = y_2 - z_{12}$ is nonnegative. But $z_{21} \leq z_{21} + y_2 = z_{12} + x_2$, so $z_{21} = z_{21} \wedge (z_{12} + x_2) \leq (z_{21} \wedge z_{12}) + (z_{21} \wedge x_2) = z_{21} \wedge x_2$; therefore $z_{21} \leq x_2$, and hence $z_{22} \geq 0$.

The last part of the next lemma is a consequence of a theorem from integration theory; since we use it again, we include a proof.

LEMMA 10.2 *If $f \in C(X)$, let $G = \{g : g \in -C \text{ and } g \geq f\}$. Then $\bar{f} = \inf\{g : g \in G\}$, G is directed (downward) by $\geq$, and $\mu(\bar{f}) = \inf\{\mu(g) : g \in G\}$ for any nonnegative measure μ on X.*

PROOF. Since $A \subset -C$, we have $\inf\{g : g \in G\} \leq \inf\{h : h \in A, h \geq f\} = \bar{f}$. On the other hand, for any $g \in G$ we have $\bar{f} \leq \inf\{h : h \in A, h \geq g\} = \bar{g} = g$. Taking the infimum on the right gives the first assertion. Since the minimum of two functions in G is a function in G, this set is directed downward; i.e., if $g_1, g_2 \in G$, there exists g in G with $g \leq g_1$, $g \leq g_2$. To prove the last assertion, let $\beta = \inf\{\mu(g) : g \in G\}$; we must show that $\mu(\bar{f}) = \beta$. Choose a sequence $\{g_n\}$ from G such that $\mu(g_n) \to \beta$. Since G is directed downward, we can assume that the sequence $\{g_n\}$ is monotonically decreasing, and hence converges pointwise to a (Borel measurable

and bounded) function f', with $f' \geq \bar{f}$. From the monotone convergence theorem, $\mu(f') = \beta$, and we will show that $\mu(f' - \bar{f}) = 0$. If
not, then $\{x : f'(x) > \bar{f}(x)\}$ has positive measure, and hence there
is a real number r and $\varepsilon > 0$ such that $\{x : \bar{f}(x) < r - \varepsilon, f'(x) > r\}$
has positive measure. This latter set contains a compact set K of
positive measure, and for each point x in K there is a function g in
G such that $g(x) < r - \varepsilon$. By compactness, we can find functions
$g'_1, g'_2, \ldots, g'_m$ in G such that for each x in K, there is a function g'_j
with $g'_j(x) < r - \varepsilon$. Let f_k in G be less than or equal to the minimum
of $g_k, g'_1, \ldots, g'_m$; then on K, we have $f_k < r - \varepsilon < f' - \varepsilon \leq g_k - \varepsilon$,
while $f_k \leq g_k$ on $X \setminus K$. Thus, $\beta \leq \mu(f_k) \leq \mu(g_k) - \varepsilon\mu(K)$ for each
k, which leads to a contradiction.

As an application of this lemma, we prove the converse to Proposition 4.2, obtaining a useful characterization of maximal measures.

PROPOSITION 10.3 (MOKOBODZKI) *A positive measure μ on X is
maximal if and only if $\mu(f) = \mu(\bar{f})$ for each continuous convex function f on X. Equivalently, μ is maximal if and only if $\mu(f) = \mu(\bar{f})$
for each $f \in C(X)$.*

PROOF. In view of Proposition 4.2, we need only show that if $\mu(f) =
\mu(\bar{f})$ for each f in C, then μ is maximal. Choose a maximal measure
λ with $\lambda \succ \mu$. Then for f in C, $\lambda(\bar{f}) = \lambda(f) \geq \mu(f) = \mu(\bar{f})$. If
$g \in -C$, then $\lambda(g) \leq \mu(g)$, so (by Lemma 10.2)

$$\lambda(\bar{f}) = \inf\{\lambda(g) : g \in -C, g \geq f\} \leq$$

$$\leq \inf\{\mu(g) : g \in -C, g \geq f\} = \mu(\bar{f}).$$

It follows that $\lambda(f) = \mu(f)$ for each f in C; since $C - C$ is dense in
$C(X)$, $\lambda = \mu$ and hence μ is maximal.

We will apply this result shortly to obtain an important fact
about the set of all maximal measures. First, however, we present
an elementary lemma concerning vector lattices. Suppose that P_1
and P_2 are cones in a vector space E, with $P_1 \subset P_2$. Denote the
induced partial orderings by $\leq_1$ and $\leq_2$, respectively. We say that
P_1 is an *hereditary subcone* of P_2 if $y \in P_1$, $x \in P_2$, and $x \leq_2 y$ imply
$x \in P_1$.

LEMMA 10.4 *If P_2 is a lattice (in the ordering $\leqq_2$) and P_1 is an hereditary subcone of P_2, then P_1 is a lattice (in the ordering $\leqq_1$).*

PROOF. Suppose that x, y are in P_1, and let z be their greatest lower bound in P_2. Then $z \leqq_2 x$, so $z \in P_1$, and we will show that if $w \leqq_1 x$ and $w \leqq_1 y$, then $w \leqq_1 z$, so that z is the greatest lower bound in P_1. Since $P_1 \subset P_2$, we see that $w \leqq_2 x$ and $w \leqq_2 y$, hence $0 \leqq_2 w \leqq_2 z$. It follows that $z - w \in P_2$ and that $z - w \leqq_2 z$; by the hereditary property, $z - w \in P_1$, so that $w \leqq_1 z$.

PROPOSITION 10.5 *The set Q of all nonnegative maximal measures on X is a subcone of the cone P of all nonnegative measures on X. The convex set $Q_1 = \{\mu : \mu \in Q, \mu(X) = 1\}$ is a base for Q, and Q_1 is a simplex.*

PROOF. Suppose that λ and μ are maximal measures; by Proposition 10.3, $(\lambda + \mu)(f) = (\lambda + \mu)(\bar{f})$ for each continuous convex function f on X, so $\lambda + \mu$ is maximal. Similarly, $r\mu$ is maximal if $\mu \in Q$ and $r \geqq 0$. Since Q_1 is the intersection of Q with the probability measures, it is clearly a base for Q. To see that Q_1 is a simplex, we must show that Q is a lattice in its natural ordering. By the previous lemma (and the fact that P is a lattice) we need only show that Q is hereditary in P. Suppose, then, that $0 \leq \lambda \leq \mu$ and that $\mu \in Q$. Let λ_1 be a maximal measure with $\lambda_1 \succ \lambda$. Then $\lambda_1 + (\mu - \lambda) \succ \lambda + (\mu - \lambda) = \mu$; since μ is maximal, we have $\mu = \lambda_1 = \mu - \lambda$, so that $\lambda = \lambda_1 \in Q$.

We now know that Q_1 is a simplex (not necessarily closed; see Section 11); furthermore, we have an affine function from Q_1 onto X defined by $\mu \to r(\mu)$, where $x = r(\mu)$ is the resultant of μ in X. (It is easily checked that this function is affine, and Lemma 4.1 asserts that it is onto.) It follows that X itself will be a simplex if this function is one-to-one, i.e., if to each x in X there exists a *unique* maximal measure μ with $\mu \sim \varepsilon_x$. This is the assertion "(5) implies (1)" of the Choquet-Meyer uniqueness theorem.

THEOREM (Choquet-Meyer). *Suppose that X is a non-empty compact convex subset of a locally convex space E. The following assertions are equivalent*

(1) X is a simplex.

(2) For each convex f in $C(X)$, the function $\bar{f}$ is affine on X.

(3) If μ is a maximal measure on X with resultant x, and if f is a convex function in $C(X)$, then $\bar{f}(x) = \mu(f)$.

(4) For each convex f and g in $C(X)$, $\overline{f+g} = \bar{f} + \bar{g}$.

(5) For each $x \in X$ there is a unique maximal measure μ_x such that $\mu_x \sim \varepsilon_x$.

To prove that (1) implies (2), we need to sharpen Proposition 3.1 a bit; recall that it asserts that for x in X and f in $C(X)$, $\bar{f}(x) = \sup\{\mu(f) : \mu \sim \varepsilon_x\}$.

LEMMA 10.6 *If $f \in C(X)$, then for each x in X, $\bar{f}(x) = \sup\{\mu(f) : \mu$ is a discrete measure and $\mu \sim \varepsilon_x\}$.*

PROOF. By a "discrete measure" we mean, of course, a measure which is a finite convex combination of measures of the form ε_y. It follows from Proposition 3.1 that we need only show the following: Given f in $C(X)$, x in X, $\mu \sim \varepsilon_x$, and $\epsilon > 0$, there exists a discrete measure λ such that $\lambda \sim \varepsilon_x$ and $\mu(f) - \lambda(f) < \epsilon$. To this end, cover X by a finite number of closed convex neighborhoods U_i such that $|f(y) - f(z)| < \epsilon/2$ whenever $y, z \in U_i \cap X$. Let $V_1 = U_1 \cap X$ and let $V_i = (U_i \cap X) \setminus (V_1 \cup \cdots \cup V_{i-1})$ for $i > 1$. Then the V_i are pairwise disjoint Borel subsets of X and for those i such that $\mu(V_i) \neq 0$ we can obtain probability measures λ_i on X, supported by V_i, by defining $\lambda_i(B) = \mu(V_i)^{-1}\mu(B \cap V_i)$ (B a Borel subset of X). Let x_i be the resultant of λ_i; since V_i is a subset of the compact convex set $U_i \cap X$, the latter must contain x_i. Define $\lambda = \Sigma\mu(V_i)\varepsilon_{x_i}$. If h is a continuous affine function on X, then $\lambda(h) = \Sigma\mu(V_i)\lambda_i(h) = \Sigma \int_{V_i} h \, d\mu = \mu(h) = h(x)$, so $\lambda \sim \varepsilon_x$. Furthermore, $\mu(f) - \lambda(f) = \Sigma[\int_{V_i} f \, d\mu - \mu(V_i)f(x_i)] = \Sigma \int_{V_i}[f - f(x_i)]d\mu < \epsilon\Sigma\mu(V_i) = \epsilon$, and the proof is complete.

Proof that (1) implies (2): Suppose that $x_1, x_2 \in X$, $\alpha_1, \alpha_2 > 0$, $\alpha_1 + \alpha_2 = 1$, and $f \in C$. Let $z = \alpha_1 x_1 + \alpha_2 x_2$; we want to show that $\bar{f}(z) = \alpha_1\bar{f}(x_1) + \alpha_2\bar{f}(x_2)$. Since $\bar{f}$ is concave, it suffices to show that it is also convex. By Lemma 10.6, we have $\bar{f}(z) = \sup\{\mu(f) : \mu$ is discrete and $\mu \sim \varepsilon_z\}$. Suppose that μ is discrete and $\mu \sim \varepsilon_z$; then there exist numbers $\beta_j \geq 0$ and a finite sequence y_j in X $(j \in J)$ such that $\Sigma\beta_j = 1$ and $\mu = \Sigma\beta_j\varepsilon_{y_j}$, i.e., $\alpha_1 x_1 + \alpha_2 x_2 = z = \Sigma\beta_j y_j$. By applying the Decomposition Lemma 10.1 to the elements $\alpha_i x_i$, $\beta_j y_j$ of $\tilde{X}$, we can choose z'_{ij} in $\tilde{X}$ such that $\alpha_i x_i = \Sigma_{j \in J} z'_{ij}$ $(i = 1, 2)$ and $\beta_j y_j = z'_{1j} + z'_{2j}$ $(j \in J)$. Each $z'_{ij} = \gamma_{ij} z_{ij}$ $(\gamma_{ij} \geq 0, z_{ij} \in X)$, and hence $x_i = \Sigma_J \alpha_i^{-1}\gamma_{ij} z_{ij}$ is a convex combination of elements of X. (Use the fact that $X \subset L^{-1}(r)$ for some L in E^*, $r \neq 0$, to see that the coefficients have sum equal to 1.) It follows that for $i = 1, 2$, the right side represents a discrete measure $\mu_i \sim \varepsilon_{x_i}$, and therefore $\bar{f}(x_i) \geq \mu_i(f) = \Sigma_j \alpha_i^{-1}\gamma_{ij} f(z_{ij})$. On the other hand, $\mu(f) = \Sigma\beta_j f(y_j)$ and for each j,

$$f(y_j) = f(\beta_j^{-1}\gamma_{1j} z_{1j} + \beta_j^{-1}\gamma_{2j} z_{2j}) \leq \beta_j^{-1}\gamma_{1j} f(z_{1j}) + \beta_j^{-1}\gamma_{2j} f(z_{2j}),$$

so $\mu(f) \leq \alpha_1\mu_1(f) + \alpha_2\mu_2(f) \leq \alpha_1\bar{f}(x_1) + \alpha_2\bar{f}(x_2)$. Taking the supremum over all discrete $\mu \sim \varepsilon_z$ gives the desired conclusion.

Proof that (2) implies (3): If μ is maximal and $f \in C$, then $\mu(f) = \mu(\bar{f})$. Since $\bar{f}$ is affine and upper semicontinuous, Lemma 10.7 (below) implies that if $\mu \sim \varepsilon_x$, then $\mu(\bar{f}) = \bar{f}(x)$.

LEMMA 10.7 *Suppose that f is an affine upper (or lower) semicontinuous function on X and that $\mu \sim \varepsilon_x$. Then $\mu(f) = f(x)$.*

PROOF. It suffices to prove that the family H of all h in A such that $h > f$ is directed downward and that $f = \inf\{h : h \in H\}$. Indeed, if this be true, then (just as in the proof of Lemma 10.2) we have $\mu(f) = \inf\{\mu(h) : h \in H\}$ for any μ; in particular, if $\mu \sim \varepsilon_x$, then $\mu(f) = \inf\{h(x) : h \in H\} = f(x)$. It remains, then, to prove the assertion about H. To see that H is directed downward, suppose that $h_1 > f$ and $h_2 > f$ $(h_i$ in $A)$; we want h in A such that $h > f$ and $h \leq h_1, h_2$. To this end, define subsets J, J_1, and J_2 of $E \times R$ as follows: $J = \{(x, r) : x \in X, r \leq f(x)\}$, $J_i = \{(x, r) : x \in X, r = h_i(x)\}$. Since f is affine and upper semicontinuous, J is closed and convex, while the continuity of h_i implies that J_i is compact. Furthermore, J is disjoint from the convex hull J_3 of $J_1 \cup J_2$, and J_3 is compact. By the separation theorem, (applied to 0 and the closed convex "difference set" $J_3 - J$) there exists a continuous linear functional L on $E \times R$ such that $\sup L(J) < \inf L(J_3) = \alpha$. The function h defined on X by $L(x, h(x)) = \alpha$ will do what is needed. A similar (but much simpler) argument shows that $f = \inf\{h : h \in H\}$. Finally, if f is affine and lower semicontinuous, we can apply what we have just proved to $-f$.

Proof that (3) implies (4): Suppose that $f, g \in C$ and that $x \in X$. Choose a maximal measure $\mu \sim \varepsilon_x$; from (3) we get $(\overline{f + g})(x) = \mu(f + g) = \mu(f) + \mu(g) = \bar{f}(x) + \bar{g}(x)$.

Proof that (4) implies (5): Suppose that $x \in X$ and consider the functional defined for f in C by $f \to \bar{f}(x)$. This is positive-homogeneous, and (4) implies that it is additive. From this it follows that $m(f - g) = \bar{f}(x) - \bar{g}(x)$ defines a linear functional m on the subspace $C - C$, and property (c) of upper envelopes (Section 3) shows that $|m(f-g)| \leq \|f-g\|$. Thus, m is uniformly continuous on the dense subspace $C - C$ of $C(X)$ and hence has a unique extension to a continuous linear functional of norm at most 1 on $C(X)$. Since $m(1) = 1$, this functional is given by a probability measure, which we denote by μ_x. Since, for f in C, we have $\mu_x(f) = m(f) = \bar{f}(x)$, Proposition 3.1 implies that $\mu_x(f) = \sup\{\mu(f) : \mu \sim \varepsilon_x\}$, i.e., $\mu_x \succ \mu$ whenever $\mu \sim \varepsilon_x$. It follows that μ_x is the unique maximal measure which represents x.

We consider next the problem of uniqueness of representing measures which are supported by the extreme points. The following easy corollary to Proposition 10.3 will enable us to prove Choquet's original uniqueness theorem for metrizable X.

COROLLARY 10.8 *If μ is a nonnegative measure on X which vanishes on every compact subset of $X \setminus \text{ex}\, X$, then μ is maximal. In particular, if μ is supported by $\text{ex}\, X$, then μ is maximal.*

PROOF. It is immediate from the hypothesis that μ vanishes on every F_σ subset of $X \setminus \text{ex}\, X$, hence is supported by every G_δ containing $\text{ex}\, X$. In particular, then, it is supported by every set of the form $\{x : f(x) = \bar{f}(x)\}$, f in $C(X)$. Thus, $\mu(f) = \mu(\bar{f})$ for f in $C(X)$, and Proposition 10.3 implies that μ is maximal.

As the example of Mokobodzki (below) will show, we cannot

weaken the hypothesis in this corollary to "μ vanishes on the compact Baire subsets of $X \setminus \mathrm{ex}\, X$."

COROLLARY 10.9 *If X is a simplex and if $\mathrm{ex}\, X$ is a Baire set, or is an F_σ set, then for each x in X there exists a unique measure μ such that $\mu \sim \varepsilon_x$ and $\mu(\mathrm{ex}\, X) = 1$.*

PROOF. By the Choquet-Meyer theorem, there exists a unique maximal measure μ such that $\mu \sim \varepsilon_x$. From Section 4 we know that $\mu(\mathrm{ex}\, X) = 1$. Suppose that $\lambda \sim \varepsilon_x$ and $\lambda(\mathrm{ex}\, X) = 1$; it follows from Corollary 10.8 that λ is maximal, and hence $\lambda = \mu$.

THEOREM (Choquet). *Suppose that X is a closed convex metrizable subset of a locally convex space. Then X is a simplex if and only if for each x in X there exists a unique measure μ which represents x and is supported by the extreme points of X.*

PROOF. It was shown in Section 3 that for metrizable X there exists a (strictly convex) continuous function f on X such that $\mathrm{ex}\, X = \{x : \bar{f}(x) = f(x)\}$. Suppose that X is a simplex; then the previous remark shows that $\mathrm{ex}\, X$ is a Baire set and Corollary 10.9 yields uniqueness. Conversely, suppose that to each x in X there corresponds a unique measure $\mu_x \sim \varepsilon_x$ with $\mu_x(\mathrm{ex}\, X) = 1$. We can conclude that X is a simplex if we show that for each x in X there is a unique *maximal* measure $\lambda \sim \varepsilon_x$. But if λ is maximal, then λ is supported by $\{x : \bar{f}(x) = f(x)\} = \mathrm{ex}\, X$, hence (by hypothesis) $\lambda = \mu_x$.

EXAMPLE (Mokobodzki)

There exists a compact convex set X with the following properties

(i) *X is a simplex.*

(ii) *$\mathrm{ex}\, X$ is a Borel set, but not a Baire set.*

(iii) *There exists a point x in X with two representing measures μ and ν such that $\mu(X \setminus \mathrm{ex}\, X) = 0$ and $\nu(X \setminus \mathrm{ex}\, X) = 1$, but ν vanishes on every Baire subset of $X \setminus \mathrm{ex}\, X$.*

PROOF. Let Y be a compact Hausdorff space containing a point y_0 which is not a G_δ, and μ a nonatomic probability measure on Y. For example, we could take Y to be an uncountable product of unit intervals, y_0 any point in Y which does not have a countable neighborhood base, and μ the corresponding product of Lebesgue measure with itself. Let $M \subset C(Y)$ be the set of all f in $C(Y)$ such that $f(y_0) = \int_Y f \, d\mu$. We first show that M separates points of Y. Suppose that λ_1 and λ_2 are probability measures on Y, with $\lambda_1(f) = \lambda_2(f)$ for all f in M. (We will soon take λ_1 and λ_2 to be point masses.) Thus, $(\lambda_1 - \lambda_2)(f) = 0$ whenever $f \in C(X)$ and $(\mu - \varepsilon_{y_0})(f) = 0$, so that these functionals are necessarily proportional, i.e., there exists a real number r such that

$$(2) \qquad\qquad \lambda_1 - \lambda_2 = r(\mu - \varepsilon_{y_0}).$$

Since μ has no atoms, it follows from (2) that λ_1 and λ_2 cannot be distinct point masses, i.e., M separates points of Y. Furthermore, if $y \neq y_0$, then y is in the Choquet boundary $B(M)$ of M. Indeed, if $\lambda \sim \varepsilon_y$, take $\lambda_1 = \lambda$, $\lambda_2 = \varepsilon_y$ in (2) and apply both sides to $\{y\}$, getting $\lambda_1(\{y\}) - 1 = 0$, so $\lambda = \varepsilon_y$. On the other hand, y_0 has two representing measures (μ and ε_{y_0}), so we can conclude that $B(M) = Y \setminus \{y_0\}$. We let $X = K(M)$; from Section 6 we know that $\mathrm{ex}\, X = \phi(B(M)) = \phi(Y) \setminus \{\phi(y_0)\}$ and that every maximal measure on X is supported by the compact set $\phi(Y)$, hence can be identified with a measure on Y. (We will use the same symbol for a measure on X which is supported by $\phi(Y)$ and for the corresponding measure on Y.) Let Q_1 be the set of all maximal probability measures on X; then a probability measure λ is in Q_1 if and only if $\lambda(\mathrm{ex}\, X) = 1$. Indeed, the latter property implies that λ is maximal, by Corollary 10.8. Suppose, on the other hand, that λ is maximal but that $\lambda(\mathrm{ex}\, X) < 1$; since λ is supported by $\phi(Y)$, we must have $\alpha = \lambda(\{y_0\}) > 0$. Let $\lambda_1 = (\lambda - \alpha\varepsilon_{y_0}) + \alpha\mu$; since each term of this sum is nonnegative, $\lambda_1 \geq 0$. Since $\mu \sim \varepsilon_{y_0}$, we have $\mu \succ \varepsilon_{y_0}$ and therefore $\lambda_1 \succ \lambda$. Clearly, $\lambda_1 \neq \lambda$, so λ is not maximal, a contradiction.

Now, we know that the resultant map r is affine from Q_1 onto X; if we show that it is one-to-one, we can conclude that X (like Q_1) is a simplex. But if $r(\lambda_1) = r(\lambda_2)$, then (by definition) $\lambda_1(f) = \lambda_2(f)$ for f in M, so that equation (2) applies to these two measures. Since each necessarily vanishes on $\{y_0\}$, we conclude that $r = 0$ and hence

$\lambda_1 = \lambda_2$. It remains, then, to prove the assertions (ii) and (iii). Let $x = \phi(y_0)$, $\nu = \varepsilon_{\phi(y_0)}$ and consider μ as a measure on X. It is clear that μ and ν represent x, that $\mu(X \setminus \mathrm{ex}\, X) = \mu(X \setminus \phi(Y)) = 0$, and that $\nu(X \setminus \mathrm{ex}\, X) = 1$. If ν were positive on a Baire subset of $X \setminus \mathrm{ex}\, X$, then it would be positive on some compact G_δ subset D of $X \setminus \mathrm{ex}\, X$, and therefore $\phi(y_0)$ would be in D. But $D \cap \phi(Y) = \{\phi(y_0)\}$ would then be a G_δ relative to $\phi(Y)$, contradicting the fact that $\{y_0\}$ is not a G_δ set. Thus, ν vanishes on every Baire subset of $X \setminus \mathrm{ex}\, X$, and if $\mathrm{ex}\, X$ were a Baire set, then $\nu(X \setminus \mathrm{ex}\, X)$ would equal zero.

We conclude this section with a proof that the definition of "simplex" coincides with the usual one for finite dimensional spaces.

PROPOSITION 10.10 *Suppose that the space $\widetilde{X} - \widetilde{X}$ spanned by X has finite dimension n. Then X is a simplex if and only if it is the convex hull of n linearly independent points. Equivalently, X has exactly n extreme points.*

PROOF. We can assume without loss of generality that $E = \widetilde{X} - \widetilde{X}$. Suppose that X has exactly n extreme points $x_1, x_2, \ldots, x_n$; since X is the convex hull of its extreme points and since X generates the n-dimensional space E, these points must be linearly independent, and hence they form a basis for E. Choose a basis $f_1, \ldots, f_n$ for E^* such that $f_i(x_j) = \delta_{ij}$. The map $T : E \to R^n$ defined by $Tx = (f_1(x), \ldots, f_n(x))$ is linear, one-to-one and onto, and carries x_i onto the i-th "unit vector" in R^n. Thus, TX is the convex hull of the n unit vectors in R^n and $\widetilde{TX}$ is the cone of nonnegative elements in R^n. This cone induces a lattice ordering in R^n, so TX is a simplex; it follows that X itself is a simplex. To finish the proof, suppose that X is a simplex and note that X is the convex hull of its extreme points, by the Minkowski theorem. Since X generates E, it must have at least n extreme points; we will show that it has *exactly* n extreme points. Suppose that the points $y_1, y_2, \ldots, y_{n+1}$ are distinct extreme points of X. Since E is n-dimensional, there exist numbers α_i, not all zero, such that $\Sigma \alpha_i y_i = 0$. Partition the integers from 1 through $n + 1$ into the sets P and N, where $i \in P$ if $\alpha_i \geq 0$, $i \in N$ otherwise. Then if $\alpha = \Sigma_{i \in P}\alpha_i$, we have $\alpha > 0$ and (since $f(X) = 1$ for some f in E^*) $\Sigma \alpha_i = 0$ so $\alpha = -\Sigma_{i \in N}\alpha_i$. Finally, let $x = \Sigma_{i \in P}\alpha^{-1}\alpha_i y_i = \Sigma_{i \in N}(-\alpha)^{-1}\alpha_i y_i$. Since these are convex combinations, we have represented an element x by two different measures on X which have support contained in $\operatorname{ex} X$. It follows from Choquet's uniqueness theorem that X is not a simplex. (This last step may be proved in a more elementary way by using the decomposition lemma and the fact that the points y_i are extreme.)

11 Properties of the resultant map

As was seen in Proposition 1.1, the resultant map from the probability measures $P(X)$ onto the compact convex set X is affine and weak* continuous. By the Choquet-Bishop-deLeeuw theorem, its restriction r to the set $Q(X)$ of maximal probability measures is still surjective, and from the uniqueness theorem we know that r is bijective if and only if X is a simplex. In this section we prove some additional properties of this map, including a simple but potentially useful selection theorem for the metrizable case.

PROPOSITION 11.1 *Suppose that the compact convex set X is a simplex. Then the inverse map $r^{-1} \colon X \to Q(X)$ exists and has the following properties:*

(i) r^{-1} is affine.

(ii) For each $f \in C(X)$ the real-valued function

$$x \to r^{-1}(x)(f) \tag{1}$$

is Borel measurable.

(iii) r^{-1} is continuous if and only if $\operatorname{ex} X$ is closed.

PROOF. (i) Since $Q(X)$ is convex and r is affine, its inverse is affine.

(ii) Assume first that f is convex; then by part (3) of the Choquet-Meyer uniqueness theorem we have $r^{-1}(x)(f) = \bar{f}(x)$, for each $x \in X$. Since the right side is upper semicontinuous, it is Borel measurable, and it follows that (1) is Borel measurable, whenever f is in $C - C$, the dense subspace of $C(X)$ spanned by the convex functions. If $f \in C(X)$ is arbitrary, it is the uniform limit of a sequence from $C - C$, so that (1) is the pointwise limit of a sequence of Borel measurable functions, hence is itself Borel measurable.

(iii) Suppose that r^{-1} is continuous and that x_0 is in the closure of ex X. Then there exists a net x_α in ex X with $x_\alpha \to x_0$, and by Proposition 1.4, $r^{-1}(x_\alpha) = \varepsilon_{x_\alpha}$. Thus, $r^{-1}(x_0) = \lim r^{-1}(x_\alpha) = \lim \varepsilon_{x_\alpha} = \varepsilon_{x_0}$. To see that $x_0 \in$ ex X, suppose that $x_0 = \frac{1}{2}(y + z)$, $y, x \in X$. The probability measure $\mu = \frac{1}{2}(\varepsilon_y + \varepsilon_z)$ represents evaluation at x_0 (hence $\mu \succ \varepsilon_{x_0}$) and there exists a maximal measure dominating μ. Since X is a simplex, that measure is $r^{-1}(x_o)$, so we have

$$\varepsilon_{x_0} = r^{-1}(x_o) \succ \mu \succ \varepsilon_{x_0}.$$

Thus, $\mu = \varepsilon_{x_0}$ and hence $y = z = x$. To prove the converse, note that if ex X is closed, then $Q(X) = P(\text{ex } X)$. It follows that $Q(X)$ is weak* compact so that r is in fact an affine homeomorphism.

Bauer [5] has given several characterizations of those X which satisfy "X is a simplex and ex X is closed" (which is why simplices with this property are often called *Bauer* simplices). Note that the foregoing proposition shows that *any Bauer simplex X can be identified with the set of all probability measures on the compact Hausdorff space* ex X.

If X is not a simplex, it is still possible to choose, for each $x \in X$, *some* maximal measure μ_x having resultant x. We present two results which give conditions under which this can be done in an affine way (Proposition 11.2) and in a measurable way (Theorem 11.4), respectively. The first of these is due to H. Fakhoury [35].

DEFINITION. *By a* selection *for the map r we mean a map $x \to \mu_x$ from X into $Q(X)$ such that $r(\mu_x) = x$ for each $x \in X$.*

PROPOSITION 11.2 *Suppose that P_1 and P_2 are cones in real vector spaces, that P_1 is lattice-ordered and that ψ is an order-preserving, additive and positive homogeneous map of P_1 onto P_2. If there exists another map ϕ from P_2 into P_1 such that $\psi \circ \phi$ is the identity on P_2, then P_2 is lattice-ordered. In particular, if there exists an affine selection for the resultant map from $P(X)$ onto X, or for its restriction $r \colon Q(X) \to X$, then X is a simplex.*

PROOF. It is straightforward to verify that if $x, y \in P_2$, then $x \vee y$ exists and is given, in fact, by $\psi[\phi(x) \vee \phi(y)]$, and this is all that is needed. To apply this result to obtain the assertion about r, say, one first extends it and its selection by homogeneity to the cones $P_1 = R^+ Q(X)$ and $P_2 = \tilde{X}$. (As in Section 10, we have assumed without loss of generality that X is contained in a hyperplane which misses the origin, so that $\tilde{X} = R^+ X$.)

The property in part (ii) of Proposition 11.1 could be more succinctly described by saying that r^{-1} is "weak* Borel measurable." More precisely, we will make use of the following terminology.

DEFINITION. *A function ϕ from a compact Hausdorff space X into a compact Hausdorff space Y is said to be* Borel measurable *provided $\phi^{-1}(U)$ is a Borel subset of X whenever U is an open subset of Y. If A is a separating family of continuous real-valued functions on Y, we will say that ϕ is* A-weakly Borel measurable *if the real-valued function $f \circ \phi$ is Borel measurable on X, for each $f \in A$*

The lemma below shows (the useful standard fact) that if Y is a metric space, then the two kinds of measurability coincide.

LEMMA 11.3 *Using the notation of the above definitions, suppose that A is a separating family of continuous real-valued functions on the compact metric space Y. Then a function $\phi: X \to Y$ is Borel measurable if (and only if) it is A-weakly Borel measurable.*

PROOF. Since A separates points of the compact space Y, the weak topology which it defines on Y coincides with the initial topolgy. Since Y is a metric space, any open subset of Y is a countable union of basic open sets of the form $\cap_{i=1}^n f_i^{-1}(I_i)$, where each $f_i \in A$ and each I_i is an open real interval. Thus, to show that $\phi^{-1}(U)$ is a Borel set whenever U is open in Y, we can assume that U has the above form. But then $\phi^{-1}(U) = \cap (f_i \circ \phi)^{-1}(I_i)$ and each set in the intersection is, by hypothesis, a Borel subset of X.

The following selection theorem was proved independently by M. Rao [67] and G. Vincent-Smith [79]; the proof below is Rao's.

THEOREM 11.4 *Suppose that X is a metrizable compact convex set. Then there exists a Borel measurable map $x \to \mu_x$ from X into the* *probability measures on* ex X *such that, for each* $x \in X$, *the measure* μ_x *represents* x, *and such that* μ_x *is an extreme point of the set of all probability measures on* ex X *which represent* x.

PROOF. For simplicity of notation in the induction argument which follows, we will consider X to be the state space of $A(X)$ and write L (rather than x) for an element of X. As in the proof of Choquet's Theorem (Section 3), the metrizability of X implies the existence of a strictly convex function f_0 in $C(X)$ as well as the existence of a dense sequence $\{f_n\}_{n=1}^{\infty}$ in $C(X) \setminus A(X)$. We can assume without loss of generality that f_n is not in the linear span A_n of $A(X) \cup \{f_0, f_1, f_2, \ldots, f_{n-1}\}$, for $n = 1, 2, 3 \ldots$. We thus have a sequence of closed subspaces

$$A(X) \equiv A_0 \subset A_1 \subset \ldots \subset A_{n-1} \subset A_n \subset \ldots \subset C(X)$$

such that A_n is the linear span of A_{n-1} and f_{n-1} and such that their union $A_\infty = \cup A_n$ is a dense subspace of $C(X)$. Let S_n denote the state space of A_n (hence $S_0 = X$), always considered in the weak* topology. Define $\phi_n \colon S_n \to S_{n+1}$ for each $n \geq 0$ as follows:

$$\phi_n(L)(g + \lambda f_n) = L(g) + \lambda \alpha_n(L), \quad L \in S_n, \quad g \in A_n, \quad \lambda \in R$$

where

$$\alpha_n(L) = \inf\{L(h) + \|h - f_n\| : h \in A_n\}.$$

Now, for each $h \in A_n$ the map $L \to L(h) + \|h - f_n\|$ is continuous on S_n, so the infimum over all such h is upper semicontinous, which implies that for fixed $g \in A_n$, $\lambda \in R$, the real-valued function

$$L \to \phi_n(L)(g + \lambda f_n)$$

is Borel measurable. It is clear that $\phi_n(L)(1) = 1$ for each L. Moreover, if $\lambda \neq 0$, then

$$\phi_n(L)(g + \lambda f_n) = L(g) + \lambda \inf\{L(h) + \|h - f_n\| : h \in A_n\} \leq$$

$$\leq L(g) + \lambda[L(-\lambda^{-1}g) + \|(-\lambda^{-1}g - f_n\|] \leq \|g + \lambda f_n\|,$$

so that $\|\phi_n(L)\| = 1$ and hence $\phi_n(L) \in S_{n+1}$. It follows from Lemma 11.3 that each ϕ_n is Borel measurable and consequently so, too, is each of the composition maps $\psi_n \colon S_0 \to S_{n+1}$ defined by

$$\psi_n = \phi_n \circ \phi_{n-1} \circ \cdots \circ \phi_1 \circ \phi_0, \quad n = 0, 1, 2, \ldots.$$

If $L \in X = S_0$ and $g \in A_n \subset S_{n+1}$, then $\psi(L) \in S_{n+1}$ and $\psi_n(L)(g) = \phi_n[\psi_{n-1}(L)](g) = \psi_{n-1}(L)(g)$, for each $n \geq 1$. By induction, we conclude that if $g \in A_{k+1}$, $L \in X$ and $n \geq k \geq 0$, then

$$\psi_k(L)(g) = \psi_n(L)(g),$$

while $\psi_n(L)(g) = L(g)$ if $g \in A_n$. Let S_∞ denote the state space of A_∞; the coherence property just shown makes it possible to define a map $\psi \colon X \to S_\infty$ as follows: If $L \in X$ and $g \in A_\infty = \cup A_n$, then $g \in A_n$ for some $n \geq 1$ and we can let $\psi(L)(g) = \psi_n(L)(g)$. Every element of S_∞ is uniformly continuous on the dense subspace A_∞, hence admits a unique extension to $C(X)$, so we can identify S_∞ with the set of probability measures on X. Since $\psi(L)(g) = L(g)$ whenever $L \in X$ and $g \in A_0$, each measure $\psi(L)$ has resultant L. It is also clear from the definition of ψ that the map $L \to \psi(L)(g)$ is Borel measurable for each $g \in A_\infty$. As in the proof of Proposition 11.1 (ii), the density of A_∞ implies that $L \to \psi(L)(g)$ is Borel measurable for each $g \in C(X)$ and Lemma 11.3 shows that $L \to \psi(L)$ is Borel measurable. For the remainder of the proof we will write μ_L in place of $\psi(L)$. To see that each μ_L is supported by $\operatorname{ex} X$, it suffices (as in the proof of Choquet's theorem) to show that $\mu_L(f_0) = \mu_L(\bar{f}_0)$. By definition,

$$\mu_L(f_0) = \psi_0(L)(f_0) = \phi_0(L)(f_0) = \alpha_0(L) =$$

$$= \inf\{L(g) + \|g - f_0\| \colon g \in A(X).$$

Now, if $g \in A(X)$, then so is $g' = g + \|g - f_0\|$; moreover, $g' \geq f_0$. Consequently

$$L(g) + \|g - f_0\| = L(g') \geq \inf\{L(h) \colon h \in A(X), h \geq f_0\} \equiv$$

$$\equiv \inf\{\mu_L(h)\colon h \in A(X), \quad h \geq f_0\},$$

hence $\mu_L(f_0)$ dominate this last expression. Now, $h \in A(X)$ and $h \geq f_0$ imply that $h \geq f_0$, so $\mu_L(h) \geq \mu_L(\bar{f}_0)$. Thus, $\mu_L(f_0) \geq \mu_L(\bar{f}_0)$; since $f_0 \leq \bar{f}_0$, the reverse inequality is obvious.

It remains to prove that μ_L is an extreme element of the set of probability measures which represent L. Suppose, then, that μ_1, μ_2 are two such measures and that $2\mu_L = \mu_1 + \mu_2$. It suffices to show that $\mu_L = \mu_1 = \mu_2$ on A_∞. By hypothesis, these functionals are equal on $A(X) = A_0$. Assuming that they are equal on A_n, we will show that they are equal on A_{n+1}, i.e., that $\mu_L(f_n) = \mu_1(f_n) = \mu_2(f_n)$. Recalling the definition of μ_L we see that

$$\begin{aligned}
\mu_L(f_n) &= \psi_n(L)(f_n) = \phi_n[\psi_{n-1}(L)](f_n) = \alpha_n[\psi_{n-1}(L)](f_n) \\
&= \inf\{\phi_{n-1}(L)(g) + \|g - f_n\|\colon g \in A_n\} \\
&= \inf\{\mu_L(g) + \|g - f_n\|\colon g \in A_n\} \\
&= \inf\{\mu_k(g) + \|g - f_n\|\colon g \in A_n\} \geq \mu_k(f_n),
\end{aligned}$$

the last inequality holding for $k = 1$ or 2 since $g + \|g - f_n\| \geq f_n$. It follows that $\mu_L(f_n) = \mu_1(f_n) = \mu_2(f_n)$ and an obvious induction argument completes the proof.

The foregoing theorem has been extended by M. Talagrand [76] in two ways. First, he showed (using a much longer proof) that there exists a selection mapping into the extreme maximal measures which is of first Baire class. He then extended this result to certain non-metrizable spaces X, retaining reasonable measurability properties.

COROLLARY 11.5 *Suppose that X is a compact metric space and that A is a closed subspace of (real or complex) $C(X)$ which contains the constants and separates points. Then there exists a Borel measurable map $x \to \mu_x$ from X to $P(X)$ such that for each $x \in X$, the measure μ_x represents evaluation at x and $\mu_x(B(A)) = 1$ (where $B(A)$ is the Choquet boundary for A).*

This last result can be improved substantially in certain important special cases. For instance, if X is a connected open subset of $\mathbf{C}^n$ and A consists of the continuous functions on the closure of X which are analytic in X, then there exists a selection $x \to \mu_x$ as above such that, for each continuous complex-valued function f on the Šilov boundary of A, the map $X \ni x \to \mu_x(f)$ is analytic. See, for instance, [80] and references therein.

12 Application to invariant and ergodic measures

Let S be a set, $\mathcal{S}$ a σ-ring of subsets of S, and $\mathcal{T}$ a family of measurable functions from S into S, i.e., for each T in $\mathcal{T}$ we have

$$T : S \to S \text{ and } T^{-1}A \in \mathcal{S} \text{ whenever } A \in \mathcal{S}.$$

A nonnegative finite measure μ on $\mathcal{S}$ is said to be *invariant* (or $\mathcal{T}$-invariant) if

$$\mu(T^{-1}A) = \mu(A) \text{ for each } T \text{ in } \mathcal{T} \text{ and } A \text{ in } \mathcal{S}.$$

There are many theorems in the literature which state that, under suitable hypotheses on S, $\mathcal{S}$ and $\mathcal{T}$, every invariant probability measure on $\mathcal{S}$ has a unique representation as an "integral average" of ergodic measures on $\mathcal{S}$ (definition below). In 1956, Choquet [17] observed that his representation theorem could be used to prove a fairly general theorem of this type. Subsequently, J. Feldman [37] gave an elementary measure theoretic description of the extreme points of the set of invariant probability measures which illuminates this result (and the generalization obtained via the Choquet-Bishop-de Leeuw theorem). We treat the measure theoretic aspect first.

Suppose that μ is a measure on $\mathcal{S}$. An element A of $\mathcal{S}$ is said to be invariant (mod μ), if $\mu(A \Delta T^{-1}A) = 0$ for each T in $\mathcal{T}$. [By $A \Delta B$ we mean the symmetric difference $(A \setminus B) \cup (B \setminus A)$.] The family of all μ-invariant sets will be denoted by $\mathcal{S}_\mu(\mathcal{T})$, or more simply by $\mathcal{S}_\mu$. It is easily seen that $\mathcal{S}_\mu$ is a sub-σ-ring of $\mathcal{S}$. We call an invariant measure μ *ergodic* if $\mu(A) = 0$ or $\mu(A) = 1$ for each A in $\mathcal{S}_\mu$. [There are other definitions of "ergodic" in the literature; ours is motived by Proposition 12.4 below. We will discuss this again at the end of the section.]

Now, the set of all invariant nonnegative finite measures on $\mathcal{S}$ forms a convex cone P, which generates the linear space $P - P$.

Furthermore, the convex set X of invariant *probability* measures is a base for P. We have, of course, defined no topology on $P - P$; we will do this later. First, we show that $P - P$ is a lattice and that the extreme points of X are the ergodic measures. To this end we prove a basic lemma (due originally to Feldman, although the present elementary proof is attributed by him to M. Sion).

LEMMA 12.1 *Suppose that μ and ν are measures on $\mathcal{S}$, that μ is invariant, and that ν is absolutely continuous with respect to μ (with $d\nu/d\mu = f$, say). Then ν is invariant if and only if $f = f \circ T$ a.e. μ for all T in $\mathcal{T}$.*

PROOF. If $f = f \circ T$ a.e. μ for all T in $\mathcal{T}$, and if $A \in \mathcal{S}$, then for each such T we have

$$
\nu(T^{-1}A) = \int_{T^{-1}A} f \, d\mu = \int_{T^{-1}A} f \circ T \, d\mu
$$

$$
= \int_A f \, d(\mu \circ T^{-1}) = \int_A f \, d\mu = \nu(A).
$$

To prove the converse, suppose $\nu \circ T^{-1} = \nu$ for some T in $\mathcal{T}$. Given a real number r, let $A = \{x : f(x) \leq r\}$, let $B = T^{-1}A \setminus A$ and let $C = A \setminus T^{-1}A$. Then $f - r > 0$ on B so $\nu(B) - r\mu(B) = \int_B (f - r) \, d\mu \geq 0$ and we have equality if and only if $\mu(B) = 0$. Moreover, $\nu(C) = \int_C f \, d\mu \leq r\mu(C)$. Now, $\nu(B) = \nu(T^{-1}A) - \nu(T^{-1}A \cap A) = \nu(A) - \nu(T^{-1}A \cap A) = \nu(C)$, and similarly, $\mu(B) = \mu(C)$. Thus, $\nu(B) \geq r \, \mu(B) = r \, \mu(C) \geq \nu(C) = \nu(B)$, so equality holds throughout. It follows that $\mu(B) = 0$ and $\mu(C) = 0$. Thus, for any r, $\{x : f(x) \leq r\}$ and $T^{-1}\{x : f(x) \leq r\} \equiv \{y : f(Ty) \leq r\}$ differ only by a set of μ-measure zero. Suppose, now, that g and h are real valued functions. Then (taking all unions over the countable dense set of rational numbers r) we have

$$
\{x : g(x) > h(x)\} = \cup\{x : g(x) > r \geq h(x)\}
$$

$$
= \cup[\{x : r \geq h(x)\} \setminus \{x : r \geq g(x)\}]
$$

By applying this identity to $g = f$, $h = f \circ T$, we see that $f \leq f \circ T$ a.e. μ, and by interchanging f and $f \circ T$ we conclude the proof.

COROLLARY 12.2 *If μ and ν are invariant measures and $\mu = \nu$ on $S_{\mu+\nu}$, then $\mu = \nu$ on S.*

PROOF. Let $f = d\mu/d(\mu + \nu)$, $g = d\nu/d(\mu + \nu)$. We will have $\mu(A) = \nu(A)$ for all A in S if $\int_A f\, d(\mu + \nu) = \int_A g\, d(\mu + \nu)$ for all such A, i.e., if $f = g$ a.e. $(\mu + \nu)$. Now f and g are S-measurable functions on S, and, in fact, they are $S_{\mu+\nu}$ measurable. Indeed, if $T \in \mathcal{T}$, then since μ, ν and $\mu + \nu$ are invariant, Lemma 12.1 implies that $f \circ T = f$ and $g \circ T = g$ a.e. $(\mu + \nu)$. If r is a real number and $I = (-\infty, r)$, then $f^{-1}(I)$ and $(f \circ T)^{-1}(I) = T^{-1}(f^{-1}(I))$ differ only by a set of $(\mu + \nu)$ measure zero (their symmetric difference is a subset of $\{x : f(x) \neq f(Tx)\}$) and hence $f^{-1}(I) \in S_{\mu+\nu}$. Thus, f (and similarly g) is $S_{\mu+\nu}$-measurable. If $A = \{x : (f - g)(x) > 0\}$, then $A \in S_{\mu+\nu}$ and hence $0 = \mu A - \nu A = \int_A (f - g) d(\mu + \nu)$; it follows that $f \leq g$ a.e. $(\mu + \nu)$ and an analogous argument shows $f \geq g$ a.e. $(\mu + \nu)$.

PROPOSITION 12.3 *The cone P of all finite nonnegative invariant measures on S is a lattice (in its own ordering).*

PROOF. In order to show that P is a lattice in its own ordering, it suffices to produce a greatest lower bound in P for any two non-negative invariant measures μ and ν. Let f and g be defined as in the proof of Corollary 12.2; we have $f = f \circ T$ and $g = g \circ T$ a.e. $(\mu + \nu)$ for all T in $\mathcal{T}$, hence $(f \wedge g) \circ T = f \wedge g$ a.e. $(\mu + \nu)$. Since the usual greatest lower bound $\mu \wedge \nu$ for two nonnegative measures is defined by $\mu \wedge \nu = (f \wedge g)(\mu + \nu)$, Lemma 12.1 implies that $\mu \wedge \nu$ is invariant. It follows easily that $\mu \wedge \nu$ is the greatest lower bound of μ and ν in the ordering induced by P, so P is a lattice.

PROPOSITION 12.4 *Suppose that μ is a member of the set X of all invariant probability measures on S. Then μ is an extreme point of X if and only if μ is ergodic.*

PROOF. Suppose that μ is an invariant probability measure and that $0 < \mu(A) < 1$ for some A in S_μ. Define

$$\mu_1(B) = \mu(B \cap A)/\mu(A) \text{ and } \mu_2(B) = \mu(B \setminus A)/[1 - \mu(A)];$$

then $\mu_1 \neq \mu$, $\mu = \mu(A)\mu_1 + (1 - \mu(A))\mu_2$, each μ_i is a probability measure, and moreover, each μ_i is invariant. [This uses the facts that μ is invariant and that $A \triangle T^{-1}A$ has μ measure zero, together with the identity

$$C_1 \cap (C_2 \triangle C_3) = (C_1 \cap C_2) \triangle (C_1 \cap C_3).]$$

To prove the converse, suppose $\mu(A) = 0$ or $\mu(A) = 1$ for each A in $\mathcal{S}_\mu$, and suppose $2\mu = \mu_1 + \mu_2$, where μ_1, μ_2 are invariant probability measures. It follows easily that $\mu_i = \mu$ on $\mathcal{S}_\mu \supset \mathcal{S}_{\mu+\mu_i}$, hence $\mu_i = \mu$ on $\mathcal{S}$, by Corollary 12.2

In order to apply the above results to obtain a representation theorem, we must define a locally convex topology on $P - P$ under which the convex set X of invariant probability measures is compact. We will use (as does Feldman) the method described by Choquet in [17].

Let S be a compact Hausdorff space, $\mathcal{S}$ the σ-algebra of Baire subsets of S and let $\mathcal{T}$ be any family of continuous maps T of S into itself. Each T is measurable; indeed, since T^{-1} carries the collection of compact G_δ subsets of S into itself, $\mathcal{S}$ is contained in the σ-algebra of all A such that $T^{-1}A \in \mathcal{S}$. The space of all finite signed Baire measures on S can be identified with the dual space $C(S)^*$ of $C(S)$. We will restrict ourselves to the weak* topology on $C(S)^*$. It is not difficult to show that for each T in $\mathcal{T}$, the induced map $\mu \to \mu \circ T^{-1}$ is a continuous linear transformation of $C(S)^*$ into itself which carries the compact convex set K of probability measures into itself. The set X of invariant probability measures is, of course, precisely the set of common fixed points for the family of affine transformations of K into itself induced by $\mathcal{T}$. Since the induced maps are continuous, X will be closed and therefore compact. We need additional hypotheses to insure that X will be nonempty. The Markov-Kakutani fixed-point theorem shows, for instance, that X will be nonempty if the family $\mathcal{T}$ (hence the set of induced maps) is commutative under the operation of composition. More generally, X will be nonempty if the semigroup generated by $\mathcal{T}$ admits a left-invariant mean (Day [23]). Once we know that X is nonempty, then we know that it has extreme points and the existence and uniqueness theorems apply to yield the following result.

THEOREM *If S is a compact Hausdorff space and T a family of continuous functions from S into S, then to each element μ of the set X of T-invariant probability Baire measures on S there exists a probability measure m on the Baire subsets of X such that*

$$\mu(f) = \int_X f \, dm \text{ for each } f \text{ in } C(S)$$

and $m(B) = 0$ for each Baire subset B of X which contains no ergodic measures. If the ergodic measures form a Baire subset or G_δ subset of X (e.g., if S is metrizable), then the measure m is unique.

It is immediate from this theorem that when S is metrizable, the set X of T-invariant measures is a simplex (necessarily metrizable). T. Downarowicz [27] has proved the interesting fact that *for any metrizable simplex K there exists a compact metric space S and a homeomorphism T of S onto itself for which the set of T-invariant probability measures on S is affinely homeomorphic to K.*

If, in the existence theorem above, the extreme points of X were closed in X, then it could be proved using the Krein-Milman theorem in place of the Choquet-Bishop-de Leeuw theorem. To see that ex X need *not* be closed, we reproduce an example due to Choquet [17]. (The existence of such an example is implied by Downarowicz's result, but the proof of the latter is far more complicated than that of the simple example below.)

EXAMPLE

Let $I = [0, 1]$ and let J be the circle, which we represent as the line $R(\bmod 1)$. Let ϕ be any continuous nonconstant function from I into R and define T from $S = I \times J$ into itself by $T(x, y) = (x, y + \phi(x))$. Then S is a compact Hausdorff space, T is a homeomorphism of S onto itself, and the extreme points of the set X of T-invariant probability measures on S do not form a closed subset of X. We will sketch a proof of this fact for the special case $\phi(x) = x$. For each $n \geq 1$, let μ_n be the measure which assigns mass n^{-1} to each of the n points (n^{-1}, kn^{-1}), $k = 0, 1, 2, \ldots, n - 1$. Then μ_n is an extreme point of X and the sequence $\{\mu_n\}$ converges in the weak* topology to Lebesgue measure μ on $\{0\} \times J$. Since $\phi(0) = 0$, every probability measure on $\{0\} \times J$ is in X, so μ is certainly not extreme in this set.

There are at least two other definitions of "ergodic measure" in the literature. One of these simply *defines* the ergodic measures to be the extreme points of the set of invariant probability measures; this, of course, requires further work if one is to relate the notion to its origins. Another definition goes as follows: An invariant probability measure μ is ergodic if $\mu(A) = 0$ or $\mu(A) = 1$ for each A in $\mathcal{S}_0 = \{A : A = T^{-1}A$ for each T in $\mathcal{T}\}$. Since $\mathcal{S}_0 \subset \mathcal{S}_\mu$, any measure ergodic in our sense is ergodic in this sense. The two notions clearly coincide if for each A in $\mathcal{S}_\mu$ there exists B in $\mathcal{S}_0$ such that $\mu(A\Delta B) = 0$. This occurs, for instance, if $\mathcal{T}$ consists of a single function T (or equals the semigroup generated by T)—simply let $B = \cap_{n=1}^{\infty} \cup_{k=n}^{\infty} T^{-k}A$. More general hypotheses on $\mathcal{T}$ which guarantee the equivalence of the two notions are given by Farrell [36, Cor. 1, Theorem 3] and Varadarajan [78, Lemma 3.3]. The following simple example, due to Farrell, shows that they are not always the same.

EXAMPLE

Let $S = [0, 1] \times [0, 1]$, let $\mathcal{S}$ be the Baire subsets of S and let $\mathcal{T} = \{T_1, T_2\}$, where $T_1(x_1, x_2) = (x_1, x_1)$, $T_2(x_1, x_2) = (x_2, x_2)$. Then T_1, T_2 are continuous maps of S onto the diagonal D of S, and $\mathcal{S}_0$ consists of S and the empty set. For any subset A of S, $(A\Delta T_i^{-1}A)\cap D$ is empty; it follows that any measure μ with support in D is invariant and $\mathcal{S}_\mu = \mathcal{S}$. (In fact, every invariant measure has support in D.) Thus, every such measure takes only the values 0 and 1 in $\mathcal{S}_0$, but the point masses on D are the only ones which are ergodic in our sense. It is interesting to note that the (noncommutative) semigroup generated by $\mathcal{T}$ is simply $\mathcal{T}$ itself.

13 A method for extending the representation theorems: Caps

The representation theorems which we dealt with in earlier sections were for elements of a compact convex set. As noted in Section 10, any such set can be regarded as a base for a closed convex cone, so these results lead in a natural way to representation theorems for the elements of a closed convex cone which admits a compact base. It is natural to wonder whether it is possible to obtain such theorems for a more general class of cones, but there seems to be no completely satisfactory result of this nature. There are, however, two lines of approach, both due to Choquet, which are of interest. One of these involves a more general notion of measure ("conical measure"), which is outlined in [19]. The other approach involves replacing the notion of "base" by that of "cap"; this makes it possible to extend the scope of the representation theorems. This section will be devoted to the latter approach. Throughout the section, we consider only *proper* cones K, i.e., $K \cap (-K) = \{0\}$.

In using the term "representation theorem" we mean, of course, more than the mere existence of measures which represent points; we require that, in some sense, these measures be supported by the extreme points. In the case of a convex cone, the only possible extreme point is the origin, and we must introduce the notion of an *extreme ray*.

DEFINITION. *A ray ρ of a convex cone K is a set of the form $R^+ x = \{\lambda x : \lambda \geqq 0\}$, where $x \in K$, $x \neq 0$. Since $R^+ x = R^+ y$ if $y = \lambda x$, $\lambda > 0$, any nonzero element of ρ may be said to generate ρ. A ray ρ of K is said to be an* extreme ray *of K if whenever $x \in \rho$ and $x = \lambda y + (1 - \lambda)z$, $(y, z \in K \quad 0 < \lambda < 1)$, then $y, z \in \rho$.*

We denote by exr K the union of the extreme rays of K; this set has the following useful descriptions, whose proofs are straightforward:

• Suppose $x \in K$; then $x \in \operatorname{exr} K$ if and only if $x = y + z$ $(y, z \in K)$ implies that $y, z \in R^+ x$.

• Suppose that $\leq$ denotes the partial ordering induced by K on the linear space $K - K$. An element x of K is in $\operatorname{exr} K$ if and only if $y = \lambda x$ (for some $\lambda \geq 0$) whenever $0 \leq y \leq x$.

• If K has a base B (so that B is a convex set with $0 \notin B$ and each ray of K intersects B in exactly one point), then ρ is an extreme ray of K if and only if ρ intersects B in an extreme point of B. Thus, $\operatorname{ex} B = B \cap \operatorname{exr} K$.

DEFINITION. *If K is a closed convex cone, a nonempty subset C of K is called a* cap *of K provided C is compact, convex, and $K \setminus C$ is convex.*

If K has a compact base B, for instance, then for any $r \geq 0$ the set $[0, r]B = \{\lambda x : 0 \leq \lambda \leq r, \ x \in B\}$ is a cap of K. If C is a cap of K such that $K = \bigcup_{n=1}^{\infty} nC$, then C is called a *universal cap* of K. (Thus, if K has a compact base B, then $C = [0, 1]B$ is a universal cap of K.) Note that any cap of K necessarily contains 0, and if K_1 is a closed subcone of K and C is a cap of K, then $C \cap K_1$ is a cap of K_1. The usefulness of this notion comes from the following fact.

PROPOSITION 13.1 *If C is a cap of the closed convex cone K and x is an extreme point of C, then x lies on an extreme ray of K.*

PROOF. Suppose that $x = \frac{1}{2}y + \frac{1}{2}z$, where $y, z \in K \setminus R^+ x$. Since $K \setminus C$ is a convex, at least one of these points, say y, is in C. Since C is compact and $y \neq 0$, $1 \leq \lambda_0 = \max\{\lambda : \lambda y \in C\} < \infty$. If $\lambda > \lambda_0$, then $\lambda y \notin C$ and [letting $z_\lambda = \lambda(2\lambda - 1)^{-1}z$] we have $x = \lambda'(\lambda y) + (1 - \lambda')z_\lambda$, where $0 < \lambda' = (2\lambda)^{-1} < 1$. Since $K \setminus C$ is convex and $\lambda y \in K \setminus C$, we must have $z_\lambda \in C$ for each $\lambda > \lambda_0$. It follows that $z_{\lambda_0} \in C$ and hence $x = \lambda_0'(\lambda_0 y) + (1 - \lambda_0')z_{\lambda_0}$ is not an extreme point of C.

This proof shows that if $y, z \in K$ and $\frac{1}{2}y + \frac{1}{2}z \in C$, then $y, z \in R^+ C$. The following useful fact is an immediate consequence of this remark: *If C is a cap of the cone K, and if $y, z \in K$ and $y + z \in R^+ C$, then $y, z \in R^+ C$.*

The above proposition leads immediately to an extension of the Krein-Milman theorem.

THEOREM (Choquet). *Suppose that K is a closed convex cone in a locally convex space and that K is the union of its caps. Then K is the closed convex hull of its extreme rays.*

At this point we could also state an integral representation theorem for the elements of such cones, but we will postpone this until we have given an alternative description of caps—a description which is extremely useful in constructing the examples which will follow.

PROPOSITION 13.2 *Suppose that K is a closed convex cone. A subset C of K is a cap of K if and only if C is compact and $C = \{x : x \in K \text{ and } p(x) \le 1\}$, where p is an extended real valued function on K with the following properties:*

(i) *p is lower semicontinuous and $0 \le p \le \infty$;*

(ii) *p is additive and positive-homogeneous.*

The cap C is universal if and only if p is finite valued.

PROOF. Suppose that C is a cap of K; then $0 \in C$ and the Minkowski (or gauge) functional p of C (defined by $p(x) = \inf\{\lambda > 0 : x \in \lambda C\}$) is nonnegative, lower semicontinuous, positive homogeneous, convex, and $C = \{x : p(x) \le 1\}$. It is easily verified that since $K \setminus C$ is convex, p is additive on R^+C (and $p = +\infty$ on $K \setminus R^+C$). The remark after the preceding proposition shows that if $y, z \in K$ and $p(y + z) < \infty$, then $p(y), p(z) < \infty$; it follows that p is additive on all of K. On the other hand, if p is a functional as described in (i) and (ii), and if $C = \{x : x \in K \text{ and } p(x) \le 1\}$ is compact, then it is easily verified that C and $K \setminus C$ are convex, so that C is a cap. Finally, the last assertion is immediate from the positive homogeneity of p and the definition of a universal cap.

In order to see how to formulate the Choquet-Bishop-de Leeuw theorem for a cone K (in a locally convex space E) which is the union of its caps, let us first see how it is formulated for a cone with a base. Suppose, then, that K has a base B and that $x \in K$, $x \ne 0$. Since any positive multiple of a base is a base, we can assume that

$x \in B$. It follows that x is the resultant of a maximal measure μ on B, and this measure is unique if B is a simplex. Now, as noted at the beginning of Section 10, we can assume that $B = \{y : y \in K$ and $f(y) = 1\}$, where f is a continuous linear functional on E. It follows that f is continuous, additive, and positive homogeneous on K, and $C = [0,1]B = \{y : y \in K, f(y) \leq 1\}$ is a cap containing x. If B is metrizable, then μ is supported by the extreme points of B; otherwise, it is supported by ex B in the sense defined in Section 4. An analogous result is valid if K is the union of its caps.

THEOREM *Suppose that K is a closed convex cone which is the union of its caps, and that $x \in K$, $x \neq 0$. Then there exists a cap $C = \{y : y \in K, p(y) \leq 1\}$ such that $x \in C_1 = \{y : y \in K, p(y) = 1\}$. Furthermore, ex $C \setminus \{0\} \subset C_1$, and any probability measure μ on C which represents x is supported by C_1. If μ is a maximal measure, then μ is supported (in an appropriate sense) by the nonzero extreme points of C.*

PROOF. Since K is the union of its caps, $x \in C = \{y : y \in K, p(y) \leq 1\}$ for some appropriate p. If $p(x) = 0$, then the compact set C would contain the ray R^+x, so we can assume $p(x) > 0$. By choosing a positive multiple of p, if necessary, we can assume $p(x) = 1$. Suppose that $y \in$ ex C, $y \neq 0$. Then $1 \geq p(y) > 0$, and $y = p(y)[y/p(y)] + [1 - p(y)] \cdot 0$, so $p(y) = 1$. If $\mu \sim \varepsilon_x$ is a probability measure on C, we can apply Lemma 10.7 to conclude that $\mu(p) = p(x) = 1$. Let $A = \{y : y \in C, p(y) < 1\}$; since p is lower semicontinuous, this is a Borel set, and if $\mu(A) > 0$, then $1 = \mu(p) = \int p \, d\mu = \int_A p \, d\mu + \int_{C_1} p \, d\mu < \mu(A) + \mu(C_1) = \mu(C) = 1$, a contradiction which shows that $\mu(C_1) = 1$. The assertion concerning maximal measures is immediate.

We now give an example which will show, among other things, that the set C_1 above need not be compact (and hence is not a base).

EXAMPLE

Let K be the convex cone of all nonnegative sequences in the space ℓ_1 of absolutely summable real sequences. Topologize K by the weak* topology induced on ℓ_1 as the dual of the space c_0 (of all real sequences which converge to 0). Then K is closed, does not have a compact base, and is not metrizable, but it has a metrizable universal cap.

Indeed, it is clear that K is closed since it is the polar set of the set of nonnegative sequences in c_0: $K = \{y = \{y_n\} : \Sigma y_n x_n \geqq 0$ whenever $x_n \to 0, x_n \geqq 0\}$. If K had a compact base B, then (as remarked at the beginning of Section 10) there would exist a weak* closed hyperplane H such that $B = H \cap K$. Thus, there would exist $z = \{z_n\} \in c_0$ with $(z, x) = \Sigma z_n x_n > 0$ for every $x \in K$, $x \neq 0$, such that $B = \{x : x \in K$ and $(z, x) = 1\}$ is compact. But the first property shows that $z_n > 0$ for all n, and hence $x^n = (0, 0, \dots, 0, z_n^{-1}, 0, 0,, \dots) \in B$. Since $z_n \to 0$, this sequence is unbounded and hence B cannot be compact. To construct a universal cap for K, define p on K by $p(x) = \Sigma x_n$. Then $C = \{x \in K : p(x) \leqq 1\}$ is compact (since it is the intersection with K of the weak* compact unit ball of ℓ_1). Since p is positive-homogeneous, $\{x \in K : p(x) \leqq r\}$ is compact for all $r > 0$, so p is lower-semicontinuous, and it is clearly additive. Since the unit ball of the dual of a separable normed linear space is always metrizable in the weak* topology, we see that C is metrizable. Finally, suppose K were metrizable. Since ℓ_1 is weak* sequentially complete and K is closed, we could conclude that K is of second category in itself. But $K = \cup nC$, and C is closed and has empty interior relative to K. (For instance, if $x \in C$, then $x + a_n \in K \setminus C$ and is weak* convergent to x, where a_n is the element of ℓ_1 which equals 2 at n and equals 0 elsewhere.)

Later, we will give an example of a cone which does not have a universal cap, but which is nevertheless the union of its caps. It is easy to construct closed cones with *no* nontrivial caps: Take a cone K generated by a base B, where B is a bounded closed convex set without extreme points. Then K has no extreme rays, hence no caps (other than $\{0\}$).

The following result gives some information concerning uniqueness of maximal measures on caps.

PROPOSITION 13.3 *If the cone K is a lattice and if C is a cap of K, then C is a simplex. Conversely, if each point of K is contained* *in a cap of K which is a simplex, then K is a lattice.*

PROOF. A cap C of a cone K is a simplex if and only if the cone C_0 generated by $C \times \{1\}$ in $E \times R$ is a lattice. If we write $C = \{x : x \in K$ and $p(x) \leq 1\}$ (for the appropriate additive functional p), then the cone C_0 may be described as follows: $x_0 = (x, r) \in C_0$ if and only if $x_0 = (0, 0)$ or $r > 0$ and $r^{-1}x_0 = (x/r, 1) \in C \times \{1\}$. This latter assertion means, of course, that $x/r \in C$, i.e., $x \in K$ and $p(x) \leq r$. It follows that $x_0 \geq 0$ if and only if $x \in K$ and $p(x) \leq r$. Assume, now, that K is a lattice; we must show that C_0 is a lattice. If $x_0 = (x, r)$ and $y_0 = (y, s)$ are in C_0, let $z = x \wedge y$ in K. Since $x - z \in K$, we have $p(x) = p(x - z) + p(z) \leq r$, so $p(z) \leq r - p(x - z)$; similarly, $p(z) \leq s - p(y - z)$. It follows that if q is the minimum of $r - p(x - z)$ and $s - p(y - z)$, then $z_0 = (z, q) \in C_0$, and we need only show that $z_0 = x_0 \wedge y_0$. It is immediate from the definition of q that $p(x - z) \leq r - q$, so $x_0 \geq z_0$; similarly, $y_0 \geq z_0$. It remains to show that if $w_0 = (w, t) \in C_0$, $x_0 \geq w_0$ and $y_0 \geq w_0$, then $z_0 \geq w_0$. The first two inequalities mean that $p(x - w) \leq r - t$ and $p(y - w) \leq s - t$. Since $p(x - w) = p(x - z) + p(z - w)$, we conclude that $p(z - w) \leq r - p(x - z) - t$; similarly, $p(z - w) \leq s - p(y - z) - t$, and hence $p(z - w) \leq q - t$, which is equivalent to $z_0 \geq w_0$.

In order to prove the partial converse, suppose that each point of K is contained in a cap which is a simplex. It suffices to show that if $x, y \in K$, then $x \wedge y \in K$. Choose a cap C of K which is a simplex and which contains the element $x + y$. As noted after the proof of Proposition 13.1, this implies that x and y are in R^+C, hence $p(x)$ and $p(y)$ are finite. Let $x_0 = (x, p(x))$, $y_0 = (y, p(y))$; then x_0, y_0 are in C_0, and by hypothesis $x_0 \wedge y_0$ exists in C_0. Denote this element by $z_0 = (z, r)$; then $z \in K$, and we will show that $z = x \wedge y$. Since $z_0 \leq x_0$ and $z_0 \leq y_0$, we have $x - z, y - z \in K$. It remains to show that if $w \in K$ and $x - w, y - w \in K$, then $z - w \in K$. Since $x = (x - w) + w \in R^+C$, we have $x - w$ and w in R^+C; similarly, $y - w \in R^+C$. If we let $w_0 = (w, p(w))$, then $x_0 \geq w_0$ and $y_0 \geq w_0$. It follows that $z_0 \geq w_0$ and hence $z - w \in K$, which completes the proof.

The remaining important question concerning caps is the following: Is there a reasonably large class of cones which are unions of their caps? This question has led Choquet [19] to investigate, in considerable depth and detail, the class of weakly complete cones. (R. Becker [7] has recently given an extensive and thorough treatment of weakly complete cones and conical measures along with numerous applications.) We restrict ourselves here to proving two results which exhibit two major classes of "well-capped" cones.

PROPOSITION 13.4 *Suppose that $K_n \subset E_n$ is a sequence of convex cones, each of which is the union of its caps. Then the same is true of any closed subcone of the product $K = \prod K_n \subset E = \prod E_n$. (In particular, any closed subcone of the countable product of R^+ with itself is the union of its caps.)*

PROOF. Since the intersection of a cap with a closed subcone of K is a cap of the subcone, we need only show that K is the union of its caps. For this, it suffices to show that if C_n is a cap of K_n, $n = 1, 2, 3, \ldots$, then there exists a cap C of K with $\prod C_n \subset C$. For each n there exists a lower semicontinuous, additive,, positive-homogeneous nonnegative functional p_n on K_n such that $C_n = \{x_n \in K_n : p_n(x_n) \leqq 1\}$. Define p on K by $p(x) = \Sigma 2^{-n} p_n(x_n)$. It is easily verified that p is also an extended real-valued nonnegative function which is additive and positive homogeneous. If ϕ_n is the projection of E onto E_n, then p is the increasing limit of the lower semicontinuous functions $\Sigma_{n=1}^N 2^{-n} p_n \circ \phi_n$, hence is lower semicontinuous. Thus, if $C = \{x \in K : p(x) \leqq 1\}$, then C is closed in E. Furthermore, if $x \in C$, then $p_n(x_n) \leqq 2^n$ for each n, so $C \subset \prod 2^n C_n$, which is compact. Finally, if $x \in \prod C_n$, then $p_n(x_n) \leqq 1$ for each n, hence $p(x) \leqq 1$ so $x \in C$ and the proof is complete.

If Y is a locally compact Hausdorff space, $C_\infty(Y)$ denotes the space (no topology) of all continuous real-valued functions on Y which have compact support. The space $M(Y)$ of all signed measures on Y which are finite on compact sets is in duality (in the obvious way) with $C_\infty(Y)$; we will consider $M(Y)$ in the weak topology induced by $C_\infty(Y)$. The following result was shown to us by P. A. Meyer.

PROPOSITION 13.5 *Suppose that Y is a locally compact, σ-compact Hausdorff space and that K is a weakly closed subcone of the cone of* all nonnegative measures in $M(Y)$. Then K is the union of its caps.

PROOF. Since Y is a countable union of compact sets, we can write $Y = \cup Y_n$, where Y_n is compact and $Y_n \subset \operatorname{int} Y_{n+1}$ for each n. Choose f_n in $C_\infty(Y)$ such that $0 \leq f_n \leq 1$, $f_n = 1$ on Y_n, $f_n = 0$ on $Y \setminus \operatorname{int} Y_{n+1}$. Suppose now, that $\mu_0 \in K$, $\mu_0 \neq 0$; we will construct a cap C which contains μ_0. Choose a sequence $\{a_n\}$ such that $a_n > 0$ and $\Sigma a_n \mu_0(f_n) = 1$. Since the sequence of nonnegative numbers $\{\mu_0(f_n)\}$ is nondecreasing (and eventually positive), the series Σa_n is convergent. For μ in K, let $p(\mu) = \Sigma a_n \mu(f_n)$. As a function on K, p is nonnegative, additive and positive-homogeneous. Let $C = \{\mu : \mu \in K \text{ and } p(\mu) \leq 1\}$. We will show that C is weakly compact and that p is lower semicontinuous; this will prove that C is a cap. To this end, define $g_n = \Sigma_{k=1}^n a_k f_k$ and let $f = \Sigma a_n f_n = \lim g_n$. The function f is strictly positive and it is continuous, since on $\operatorname{int} Y_{n+1}$, $f = g_n + \Sigma_{k=n_1}^\infty a_k$. Furthermore, if $g \in C_\infty(Y)$, then $g = 0$ in $Y \setminus \operatorname{int} Y_{n+1}$ for some n, so there exists a number $b(g) > 0$ such that $|g| \leq b(g)f$. If $\mu \in K$, then (since $g_n \nearrow f$) we have $\mu(f) = \lim \mu(g_n) = \lim \Sigma_{k=1}^n a_k \mu(f_k) = p(\mu)$. Thus, if $\mu \in C$, then for any g in $C_\infty(Y)$ we have $\pm\mu(g) \leq b(g)\mu(f) \leq b(g)$. It follows that $C \subset P = \prod\{[-b(g), b(g)] : g \in C_\infty(Y)\}$; since P is compact in the product topology (which coincides with the weak topology on C) it suffices to show that C is closed in P. It is immediate that any element in the pointwise closure of C is a nonnegative linear functional on $C_\infty(Y)$, and hence is a measure. Thus, we need only show that C is closed in K. This follows from the fact that p is the increasing limit of the continuous functions on K defined by $\mu \to \mu(g_n)$ and hence is lower semicontinuous.

We now exhibit an example of a cone which is the union of its caps, but does not have a universal cap.

EXAMPLE

Let s be the space of all real sequences in the product topology and let $E = s^*$ be the dual space of s. As is well known, E can be considered to be the space of all finitely nonzero sequences, with the correspondence defined by $(a, x) = \Sigma a_n x_n$, $a \in s$, $x \in E$. Topologize E by the weak* topology defined by s and let K be the closed convex cone of all nonnegative elements of E. If $x \in K$, we can define a cap C containing x as follows: Let $I = \{k : x_k = 0\}$, $J = \{k : x_k > 0\}$, and suppose that J has n elements. Let C be those y in K such that $y_k = 0$ for k in I and $\Sigma_{k \in J}\, y_k x_k^{-1} \leq n$. It is straightforward to verify that C is convex, $K \setminus C$ is convex and $x \in C$. Furthermore, C is a subset of the finite dimensional subspace of E consisting of all y such that $y = 0$ on I. If $y \in C$, then $0 \leq y_k \leq n x_k$ for $k \in J$; it follows that C is bounded (and closed) hence compact.

To see that K does not have a universal cap, suppose that $C = \{x : x \in K, p(x) \leq 1\}$ were such an object, for a suitable function p. Denote by δ_n the sequence which is 0 except in the n-th place, where it is 1. Since C is universal, $p(\delta_n) < \infty$, and since C is compact, $p(\delta_n) > 0$. Let $a = \{np(\delta_n)\} \in s$ and $x^n = p(\delta_n)^{-1}\delta_n \in C$; then $(a, x^n) = n$, so C is not weak* compact, a contradiction.

We conclude this section with a result which gives a topological criterion for a cone to have a compact base.

PROPOSITION 13.6 *Suppose that K is a closed convex cone in a locally convex space E such that $K \cap (-K) = \{0\}$. Then K has a compact base if and only if K is locally compact.*

PROOF. If K has a compact base B, then we can assume (as in Section 10) that $B = K \cap \{x : f(x) = 1\}$ for some f in E^* such that $f(x) > 0$ for $x \neq 0$ in K. The sets $[0, n]B = K \cap \{x : f(x) \leq n\}$, $n = 1, 2, 3, \ldots$ are compact, have nonempty interior (relative to K), and their union is K; it follows that K is locally compact. On the other hand, suppose that K is locally compact. Then there exists a convex neighborhood U of 0 such that $U \cap K$ is compact. Let F be the intersection of K with the boundary of U and let J be the closed convex hull of F. Since $U \cap K$ is compact (and convex), the same is true of J. By Milman's theorem (see Section 1), the extreme points of J are contained in F; since $J \subset K$ and $0 \in \text{ex}\, K$, we conclude that $0 \notin J$. Thus, there exists a continuous linear functional f on E which strictly separates 0 from J, i.e., $2b = \inf f(J) > 0$. It follows that $B = f^{-1}(b) \cap K$ is a base for K and is compact, since it is contained in $[0, 1]J$.

14 A different method for extending the representation theorems

When we say that a probability measure μ on a compact convex set X "represents" a point x of X, we mean, of course, that $\mu(f) = f(x)$ for each continuous affine function f on X. One way of extending the representation theorems would be to show that this latter equality holds for a larger class of functions. For instance, Proposition 10.7 showed that it holds for upper semicontinuous (or lower semicontinuous) affine functions. In this section we will show that it holds for the affine functions of *first Baire class*, i.e., those affine functions which are the pointwise limit of a sequence of continuous (but not necessarily affine) functions on X.

THEOREM (Choquet [18]). *If X is a compact convex subset of a locally convex space E and if μ is a probability measure on X with resultant x, then $\mu(f) = f(x)$ for each affine function f of first Baire class on X.*

The proof which follows uses a weaker property than that stated in the hypotheses, namely, we need only assume the following:

(1)   The function f is affine, Borel measurable, and the restriction of f to any compact subset of X has at least one point of continuity.

If f is the limit of a sequence of continuous functions, it is certainly Borel measurable, and its restriction to any compact subset of X is again of first Baire class. A classical consequence of the Baire category theorem asserts that a function of first Baire class has a dense set of points of continuity so (1) follows from the original hypotheses on f. In order to know that f is integrable with respect to μ, it suffices to prove that if a function f satisfies (1), then it is bounded. Let y be a point of continuity of f, and suppose that f is not bounded. Since X is compact, we can find a net x_α

and a point x in X such that $x_\alpha \to x$ and $\{f(x_\alpha)\}$ is unbounded. Choose an open neighborhood U of y such that f is bounded on U and choose $0 < t < 1$ such that $ty + (1 - t)x \in U$. Eventually, $u_\alpha = ty + (1 - t)x_\alpha \in U$. Since $f(u_\alpha) = tf(y) + (1 - t)f(x_\alpha)$, this leads to a contradiction.

We next introduce some notation for the *oscillation* of f: If $A \subset X$, let $Of(A) = \sup f(A) - \inf f(A)$, and for x in X, let $O_x f = \inf\{Of(U) : U \text{ open}, x \in U\}$.

LEMMA 14.1 *If μ is a nonnegative measure on X and $\varepsilon > 0$, then there exists a sequence $\{\lambda_n\}$ of nonnegative measures on X, supported by pairwise disjoint Borel subsets S_n of X, such that $\mu = \Sigma\lambda_n$ and $Of(K_n) < \varepsilon$ for each n, where K_n is the closed convex hull of S_n.*

PROOF. We will show below that if ν is any nonnegative measure on X, $\nu \neq 0$, then there is a Borel set B of positive ν measure such that $Of(K) < \varepsilon$, where K is the closed convex hull of B. Assume, for the moment, that this "induction step" has been carried out. Let Z be the collection of all sets M of nonnegative measures on X with the following three properties:

(i) Each λ in M is the restriction of μ to a Borel subset S_λ of positive μ measure.

(ii) If K_λ is the closed convex hull of S_λ, then $Of(K_\lambda) < \varepsilon$.

(iii) If $\lambda, \lambda' \in M$ and $\lambda \neq \lambda'$, then $S_\lambda, S_{\lambda'}$ are disjoint.

The collection Z is nonempty: In the induction step, simply take ν to be μ and let $M = \{\lambda\}$, where λ is the restriction of μ to B and $S_\lambda = B$. If we partially order Z by inclusion then it is easily verified that Z is an "inductive" partially ordered set, so Zorn's lemma is applicable and there necessarily exists a maximal element M_0 in Z. Since the sets S_λ ($\lambda \in M_0$) are pairwise disjoint and of positive μ measure, the set M_0 is countable; say $M_0 = \{\lambda_n\}$. It follows from the Lebesgue dominated convergence theorem, say, that the series $\Sigma\lambda_n$ converges to the restriction λ of μ to $\cup S_n$, where $S_n = S_{\lambda_n}$. If $\mu \neq \lambda = \Sigma\lambda_n$, then we can apply the induction step (above) to

$\nu = \mu - \lambda$, obtaining a Borel set B of positive μ measure (hence of positive μ measure) such that $Of(K) < \varepsilon$, where K is the closed convex hull of B. Since we can certainly assume that B is disjoint from $\cup S_n$, we are led to a contradiction of the maximality of M_0.

It remains, then, to prove the induction step. Given ν, f and $\varepsilon > 0$, let S be the closed support of ν (i.e., the complement of the union of all open sets of ν measure zero) and let J be the closed convex hull of S. Denote the restriction of f to J by g, and let $Y = \{x : x \in J, O_x g \geqq \varepsilon\}$. The set Y is closed (for any real valued function g) and from the fact that g is affine it follows that Y is convex. Since g has at least one point of continuity in J, the set $J \setminus Y$ is nonempty. If $S \subset Y$, then we would have $J \subset Y$; consequently, $S \setminus Y$ is nonempty. From the definition of S it follows that any neighborhood of any point of $S \setminus Y$ has positive measure. Since Y is closed and E is locally convex, we can choose a closed convex neighborhood of some point of $S \setminus Y$ which misses Y, and hence there exists a compact convex subset V of $J \setminus Y$ of positive ν measure; clearly, $O_x g < \varepsilon$ for x in V. Now, since f is bounded, the function g is bounded and hence we can write V as a finite union of convex sets V_k for which $Og(V_k) < \varepsilon$. (For instance, finitely many sets of the form $\{x : x \in V, (n-1)\varepsilon \leqq 2g(x) < n\varepsilon\}$, n an integer, will cover V.) At least one of these sets V_k has positive ν measure, and by regularity it must contain a compact set J_0 of positive ν measure. Let K be the closed convex hull of J_0; clearly $K \subset V$, so $O_x g < \varepsilon$ for x in K. In fact, to complete the proof we need only show that $3\varepsilon \geqq Og(K)[= Of(K)]$. Let J_1 be the convex hull of J_0; then $J_1 \subset V_k$, so $Og(J_1) < \varepsilon$. If x, y are in K (i.e., in the closure of J_1), then there exist neighborhoods (in J) U_x, U_y of x, y, respectively, for which $Og(U_x) < \varepsilon$, $Og(U_y) < \varepsilon$. It follows from the triangle inequality that $|g(x) - g(y)| < 3\varepsilon$, and the proof of the lemma is complete.

We now finish the proof of the theorem. Suppose that $\mu \sim \varepsilon_x$, that f satisfies (1), and that $\varepsilon > 0$. By the lemma, we can choose measures $\mu_1, \mu_2, \ldots, \mu_n$ and λ with disjoint supports such that $\|\lambda\| < \varepsilon$, $\mu = \Sigma \mu_k + \lambda$ and the support of μ_k is contained in a compact convex set K_k for which $(Of)(K_k) < \varepsilon$. Let $\lambda_k = \mu_k / \|\mu_k\|$ and let x_k be the resultant of λ_k. It follows that $x_k \in K_k$ and hence $f(x_k) - \varepsilon \leqq$

$\lambda_k(f) \leqq f(x_k) + \varepsilon$ for each k. Thus, $|\mu_k(f) - \|\mu_k\| f(x_k)| \leqq \varepsilon \|\mu_k\|$ for each k. Let y be the resultant of $\lambda / \|\lambda\|$; since $\mu = \Sigma \|\mu_k\| \lambda_k + \|\lambda\|(\lambda / \|\lambda\|)$ and $1 = \|\mu\| = \Sigma \|\mu_k\| + \|\lambda\|$, we have $x = \Sigma \|\mu_k\| x_k + \|\lambda\| y$, so that $f(x) = \Sigma \|\mu_k\| f(x_k) + \|\lambda\| f(y)$. Thus,

$$
\begin{aligned}
|\mu(f) - f(x)| &= |\Sigma[\mu_k(f) - \|\mu_k\| f(x_k)] + \lambda(f) - \|\lambda\| f(y)| \\
&\leqq \varepsilon \Sigma \|\mu_k\| + |\lambda(f) - \|\lambda\| f(y)| \\
&\leqq \varepsilon + 2\|\lambda\| \sup |f| < \varepsilon(1 + 2 \sup |f|).
\end{aligned}
$$

Since this holds for each $\varepsilon > 0$, $\mu(f) = f(x)$.

Choquet [18] has given an example which shows that the above theorem fails for an affine function of *second* Baire class (i.e., the pointwise limit of a sequence of functions of first Baire class). we will describe the example, but omit the proof that the function is of second Baire class. A proof may be found in [1, p. 20] .

EXAMPLE

Let X be the compact convex set of all probability Borel measures μ on $[0, 1]$. Each measure μ in X admits the unique decomposition into its purely atomic and atom–free parts, and we let $f(\mu)$ be the norm of the atomic part of μ. (Equivalently, $f(\mu) = \sum_{x \in [0,1]} \mu(\{x\})$.) It is easily checked that f is a bounded affine function on X. If μ is an extreme point of X, then μ is a point mass and $f(\mu) = 1$; consequently, $f = 1$ on the image in X of $[0, 1]$. Let λ be Lebesgue measure on $[0, 1]$; then $f(\lambda) = 0$. On the other hand, λ can be carried to a measure ν on X; then ν is supported by the extreme points of X and its resultant in X is λ. But $\nu(f) = 1 \neq f(\lambda)$.

Using Choquet's theorem, G. Mokobodzki [69, Appendice] has given an elementary proof of the fact that *every affine function of first Baire class on a compact convex set X is, in fact, the pointwise limit of a sequence of continuous affine functions on X.*

15 Orderings and dilations of measures

If X is a compact convex subset of a locally convex space E, and if μ, λ are nonnegative measures on X, we have defined $\mu \succ \lambda$ to mean that $\mu(f) \geq \lambda(f)$ for each continuous convex function f on X. For finite dimensional spaces E, this ordering has long been of interest in statistics; it is used to defined "comparison of experiments." A characterization in terms of *dilations* (defined below) has been given by Hardy, Littlewood, and Polya for one dimensional spaces, and by Blackwell [10], C. Stein, and S. Sherman for finite dimensional spaces. The general case has been proved by P. Cartier [15], based in part on the work of Fell and Meyer; this is the proof we present below. An entirely different approach has been carried out by Strassen [75].

There is another ordering, denoted by $\mu \gg \lambda$ which was introduced by Loomis [56] in the course of his proof of the Choquet-Meyer uniqueness theorem, and which is of interest in connection with the theory of group representations. The second main result of this section is the proof that $\mu \gg \lambda$ if and only if $\mu \succ \lambda$.

We will let P_1 denote the set of all regular Borel probability measures on X. A mapping $x \to T_x$ from X into P_1 is called a *dilation*

(1) The measure T_x represents x, for each x in X.

(2) For each f in $C(X)$, the real valued function $x \to T_x(f)$ is Borel measurable.

(In the language of Section 11, T is a Borel measurable selection for the resultant map.)

There is a natural extension of T to a map from P_1 into P_1, defined as follows: If $\lambda \in P_1$, let $T\lambda$ be the measure obtained (via the Riesz theorem) from the bounded linear functional defined by

$$(*) \qquad (T\lambda)(f) = \int_X T_x(f) d\lambda(x), \qquad f \in C(X).$$

Since $T_y \sim \varepsilon_y$ for all y, taking $\lambda = \varepsilon_x$ in $(*)$ shows that $T(\varepsilon_x) = T_x$, so that (modulo the homeomorphism $x \to \varepsilon_x$) this is a genuine extension; it, too, is called a dilation. We can picture the measure T_x as "spreading out" or "dilating" the unit mass at x. Condition (2) says that this should be done in a reasonable way as we change from one point to another, and $(*)$ says that $T\lambda$ is the measure obtained by taking the λ-average of these individual dilations. It is not surprising, then, that $T\lambda$ should have its support "closer" to the extreme points of X than does λ:

$$\text{If } \mu = T\lambda, \text{ then } \mu \succ \lambda.$$

Indeed, suppose that f is a continuous convex function X. Since $T_x \sim \varepsilon_x$, we have $T_x \succ \varepsilon_x$, so that $T_x(f) \geq f(x)$ for all x. It is immediate from $(*)$ that $\mu(f) = (T\lambda)(f) \geq \int f d\lambda = \lambda(f)$. The main result of this section is the following theorem of Hardy-Littlewood-Polya-Blackwell-Stein-Sherman-Cartier , under the hypothesis that X be metrizable.

THEOREM *Suppose that X is a compact metrizable convex subset of a locally convex space and that μ and λ are regular Borel probability measures on X. Then $\mu \succ \lambda$ if and only if there exist a dilation T such that $\mu = T\lambda$.*

The proof of this theorem depends on a general result of Cartier (which does not use metrizability), together with a classical result on the disintegration of measures.

With X as above, we consider the space $F = C(X)^* \times C(X)^*$, using the product of the weak* topology with itself. Thus, F is a locally convex space, and every continuous linear functional L on F is of the form

$$L(\alpha, \beta) = \alpha(f) - \beta(g), \qquad (\alpha, \beta) \in F$$

for some pair of functions f, g in $C(X)$. Throughout this section we will be interested in two particular subsets J and K of F, defined as follows:

$$K = \{(\lambda, \mu) : \lambda \geq 0, \mu \geq 0 \text{ and } \mu \succ \lambda\}$$
$$J = \{(\varepsilon_x, \nu) : x \in X, \nu \sim \varepsilon_x\}.$$

It is easily verified that K is a closed convex cone in F. Since $\nu \sim \varepsilon_x$ implies $\nu \succ \varepsilon_x$, we see that $J \subset K$; furthermore, J is compact (since the map $\nu \to$ (resultant of ν) is continuous from P_1 into X, and J is homeomorphic to its graph). Since a convex combination of point masses is not a point mass the set J is not convex. *Its closed convex hull B, however, is a compact base for K.* Indeed, J is a subset of the intersection $K \cap H$ of K with the hyperplane H of all (α, β) for which $\alpha(1) = 1$. Since $(\alpha, \beta) \in K$ and $\alpha(1) = 1$ imply $\beta(1) = 1$, we see that $K \cap H$ is a closed convex subset of the compact convex set $P_1 \times P_1$, hence is compact. Thus $B \subset K \cap H$ and is itself compact. It is clear that $K \cap H$ is a base for K; we will show that $B = K \cap H$. This will certainly be true if B generates K, i.e., if $L \geq 0$ on K whenever $L \in F^*$ and $L \geq 0$ on B. Now, if $L \geq 0$ on B, then $L \geq 0$ on J, so assume there exist f, g in $C(X)$ such that $L(\varepsilon_x, \nu) = f(x) - \nu(g) \geq 0$ whenever $\nu \sim \varepsilon_x$; we will show that $L(\alpha, \beta) = \alpha(f) - \beta(g) \geq 0$ whenever $(\alpha, \beta) \in K$. Recall (Proposition 3.1) that for each x in X, $\bar{g}(x) = \sup\{\nu(g) : \nu \sim \varepsilon_x\}$. It follows that $\bar{g}(x) \leq f(x)$, so that $g \leq \bar{g} \leq f$. Thus, $\beta(g) \leq \beta(\bar{g})$ and $\alpha(\bar{g}) \leq \alpha(f)$; from Lemma 10.2 we know that $\beta(\bar{g}) \leq \alpha(\bar{g})$ and hence $L(\alpha, \beta) \geq 0$.

The following proposition is now an immediate consequence of Proposition 1.2

PROPOSITION 15.1 (CARTIER) *An element (λ, μ) of F is in K if and only if there exists a nonnegative measure on J which represents (λ, μ).*

We now return to the proof of the theorem itself. Assume, then, that X is metrizable and that $\mu \succ \lambda$. By the above proposition, there exists a nonnegative measure m' on J such that $\int_J L \, dm' = L(\lambda, \mu)$ for each L in F^*. This means that for each (f, g) in $C(X) \times C(X)$,

$$\lambda(f) - \mu(g) = \int_J [f(x) - \nu(g)] \, dm'(\varepsilon_x, \nu).$$

Let $S = \{(x, \nu) : x \in X, \nu \in P_1, \nu \sim \varepsilon_x\}$. Since the function $(\varepsilon_x, \nu) \to (x, \nu)$ from J onto S is a homeomorphism, we can carry m' to a measure m on S. By alternately choosing $g = 0$, $f = 0$ in the above equation, we see that for all f, g in $C(X)$,

$$\begin{array}{llll}
\text{(a)} & \lambda(f) & = & \int_S f(x) \, dm(x, \nu), \\
\text{(b)} & \mu(g) & = & \int_S \nu(g) \, dm(x, \nu).
\end{array}$$

Equation (a) shows that m is a probability measure on S which is carried onto λ under the natural projection of $X \times P_1$ onto X.

We now state a special case of the theorem on disintegration of measures [13, p. 58].

Suppose that Y and X are compact metrizable spaces, that ϕ is a continuous function from Y onto X, and that m is a nonnegative measure on Y. Let $\lambda = m \circ \phi^{-1}$ denote the image of m under the function ϕ. Then there exists a function $x \to \lambda_x$ from X into the probability measures on Y, with the following properties:

(i) *For each h in $C(Y)$, the function $x \to \lambda_x(h)$ is Borel measurable.*

(ii) *For each x in X, the support of λ_x is contained in $\phi^{-1}(x)$.*

(iii) *For each h in $C(Y)$, $m(h) = \int_X \lambda_x(h) \, d\lambda(x)$.*

We apply this result as follows: Let $Y = S \subset X \times P_1$, let ϕ be the natural projection of S onto X, and let m and λ be the measures introduced previously. Then, as we have noted, (a) implies that $\lambda = m \circ \phi^{-1}$, so there exists $x \to \lambda_x$ from X into the probability measures on S, satisfying the above three properties. We let T_x be the resultant in P_1 of the image of λ_x under the natural projection of S onto P_1. It remains to prove that T_x satisfies the properties (1) and (2) which define dilations, and that $\mu = T\lambda$. The fact that T_x is the resultant in P_1 of the image of λ_x means that for each f in $C(X)$,

(**)
$$T_x(f) = \int_S \nu(f) d\lambda_x(y, \nu),$$

Since (y, ν) in S implies $\nu \sim \varepsilon_y$, we see that for continuous *affine* functions f, this becomes

$$T_x(f) = \int_S f(y)d\lambda_x(y, \nu).$$

We know that λ_x is supported by $\phi^{-1}(x) = \{(x, \nu) : \nu \sim \varepsilon_x\}$, and hence $T_x(f) = f(x)$, i.e., T_x represents x. Property (2) of dilations follows from (**) and property (i). Finally, to show that $\mu = T\lambda$, we must verify that for g in $C(X)$,

$$\mu(g) = (T\lambda)(g) = \int_X T_x(g)d\lambda(x).$$

By (**), $T_x(g) = \int_S \nu(g)d\lambda_x(y, \nu)$. Since $h(y, \nu) = \nu(g)$ defines a function h in $C(S)$, (iii) implies that

$$\begin{aligned}
\int_S \nu(g)dm(y, \nu) &= \int_X (\int_S \nu(g)d\lambda_x(y, \nu))d\lambda(x) \\
&= \int_X T_x(g)d\lambda(x).
\end{aligned}$$

From (b), we see that the left side equals $\mu(g)$, and the proof is complete.

We next define the ordering $\mu \gg \lambda$ of Loomis [56]. (Actually, Loomis considers several orderings; the present one is his "strong" ordering.)

DEFINITION. *If μ is a nonnegative measure on X, a* subdivision *of μ is a finite set $\{\mu_i\}$ of nonnegative measures on X such that $\mu = \Sigma\mu_i$. We say that $\mu \gg \lambda$ if for each subdivision $\{\lambda_i\}$ of λ there exists a subdivision $\{\mu_i\}$ of μ such that $\mu_i \sim \lambda_i$ for each i.*

(For other descriptions of this ordering and its relation to group representations, see [57] and [56].)

In the following theorem, X and J are the same as in Proposition 15.1 of Cartier. Note that X is *not* assumed to be metrizable.

THEOREM (Cartier-Fell-Meyer [15]). *If λ and μ are nonnegative measures on X, then the following assertions are equivalent:*

(a) $\mu \succ \lambda$.

(b) *There exists a nonnegative measure m on J which represents* (λ, μ).

(c) $\mu \gg \lambda$.

PROOF. Proposition 15.1 shows that (a) implies (b). Suppose that (b) holds, and let $\{\lambda_i\}$ be any subdivision of λ. By means of the Radon-Nikodym theorem we can choose nonnegative Borel measurable functions $\{f_i\}$ on X such that $\lambda_i = f_i \lambda$ and $\Sigma f_i = 1$. Define Borel functions $\{g_i\}$ on J by $g_i(\varepsilon_x, \nu) = f_i(x)$ for each (ε_x, ν) in J, and let $m_i = g_i m$. By Proposition 15.1 again, each measure m_i has a resultant (ν_i, μ_i) in the cone K. If we use the definition of this assertion (and if we carry the measure m to the set S defined after Proposition 15.1) we see that

$$\nu_i(f) = \int_S f(x) f_i(x) \, dm(x, \nu), \text{ for } f \text{ in } C(X).$$

Similarly, since m represents (λ, μ), we deduce that

$$\lambda(f) = \int_S f(x) \, dm(x, \nu), \text{ for } f \text{ in } C(X).$$

As we noted earlier, this means that $\lambda = m \circ \pi^{-1}$, where π is the natural projection of $S \subset X \times P_1$ onto X. Since the f_i are bounded Borel functions, it follows that $\lambda(f f_i) = (m \circ \pi^{-1})(f f_i)$ for each f in $C(X)$, $i = 1, 2, \ldots, n$. Now, for each f in $C(X)$ and each i, we have

$$\int_S (f f_i \circ \pi)(x, \nu) \, dm(x, \nu) = \int_X (f f_i)(x) \, d(m \circ \pi^{-1})(x),$$

so that $\nu_i(f) = \int_S f f_i dm = \lambda(f f_i) = \lambda_i(f)$, i.e., $\nu_i = \lambda_i$. But $(\lambda_i, \mu_i) \in K$ implies $\mu_i \sim \lambda_i$, and $m = \Sigma m_i$ implies $\mu = \Sigma \mu_i$, so $\mu \gg \lambda$.

It remains to show that (c) implies (a). Suppose, then, that $\mu \gg \lambda$ and that f is a continuous convex function on X; we want to show that $\mu(f) \geqq \lambda(f)$.

Given $\varepsilon > 0$, we can carry out the same construction as was used in the proof of Lemma 10.6 to write X as a disjoint union of Borel sets $V_1, V_2, \ldots, V_n$ such that the restriction λ_i of λ to V_i is nonzero and, letting x_i be the resultant in X of the probability measure $\lambda_i/\lambda_i(X)$, we will have $|f(x) - f(x_i)| < \varepsilon$ for each x in V_i. Thus, $\lambda = \Sigma \lambda_i$, and therefore we can choose measures μ_i such that $\mu = \Sigma\mu_i$ and $\mu_i \sim \lambda_i$. The latter implies that $\mu_i(X) = \lambda_i(X) = \lambda_i(V_i)$ and that x_i is the resultant of $\mu_i/\mu_i(X)$. Since f is convex, $\mu_i(f)/\mu_i(X) \geq f(x_i)$, and consequently $\mu(f) = \Sigma\mu_i(f) \geq \Sigma\lambda_i(V_i)f(x_i)$. On the other hand, $f \leq f(x_i) + \varepsilon$ on V_i, so that $\lambda_i(f) \leq \lambda(V_i)\,[f(x_i) + \varepsilon]$, and hence

$$\lambda(f) = \Sigma\lambda_i(f) \leq \Sigma\lambda_i(V_i)f(x_i) + \varepsilon\lambda(X) \leq \mu(f) + \varepsilon\lambda(X).$$

Since this is true for any $\varepsilon > 0$, we conclude that $\mu \succ \lambda_i$ and the proof is complete.

We conclude this section with an interesting proposition concerning dilations and maximal measures.

PROPOSITION 15.2 (MEYER [57]) *Suppose that X is metrizable, that λ is a nonnegative measure on X, and that μ is a maximal measure, with $\mu \succ \lambda$. Let T be a dilation such that $T\lambda = \mu$. Then T_x is maximal, almost everywhere with respect to λ.*

PROOF. Recall from Proposition 10.3 that a measure μ is maximal if and only if $\mu(f) = \mu(\overline{f})$ for every f in $C(X)$. Let $\{f_n\}$ be a countable dense subset of $C(X)$; then for each n, $0 = \mu(\overline{f}_n - f_n) = \int_X T_x(\overline{f}_n - f_n)d\lambda(x)$. Now, $\overline{f}_n - f_n \geq 0$, so we have $T_x(\overline{f}_n - f_n) = 0$ a.e. λ, for each n. It follows that for all n, $T_x(f_n) = T_x(f_n)$ a.e. λ. Since (Section 3) the map $f \to \overline{f}$ is uniformly continuous, we conclude that for almost all x, $T_x(f) = T_x(\overline{f})$ for each f in $C(X)$, and hence T_x is maximal a.e. λ.

Much of the material in these notes (other than the applications) is contained in the outline presented by Choquet [19] at the 1962 International Congress of Mathematicians, and the paper [22] by Choquet and Meyer gives an elegant and very concise treatment of the main parts of the theory. Bauer's lecture notes [6] contain a detailed development which starts from the very beginning, using (as do Choquet and Meyer) his "potential theoretic" approach to the existence of extreme points via semi–continuous functions on a compact space [3]. Chapter XI of Meyer's book [57] covers a great deal of ground. He shows, among other things, that the entire subject of maximal measures may be viewed as a special case of an abstract "theory of balayage."

A number of books and monographs on this subject have appeared since the 1966 first edition of these notes (which appeared in Russian translation in 1968 [63]). Among these have been Gustave Choquet [20] (1969), Erik M. Alfsen [1] (1971), Yu. A. Šaškin [73] (1973), S. S. Kutateladze [53] (1975), L. Asimow and A. J. Ellis [2] (1980) and Phelps [61] (1980). In [21], Choquet has given a survey (without proofs) of related results obtained through 1982.

The 245–page book by R. Becker [7] (1999) contains a superb up-to-date exposition which in many respects starts where these notes leave off. His emphasis on convex cones (rather than compact convex sets) and conical measures permits applications to potential theory, capacities, statistical decision theory and other topics where the cone of interest does not admit a compact base.

The 474–page monograph by Bourgin [14] is extraordinarily thorough; in particular, his Chapter 6 covers integral representations for elements of certain non–compact convex subsets of Banach spaces (sets with the RNP; see below).

In [83], G. Winkler has focussed on the Choquet ordering and noncompact convex sets (not necessarily in Banach spaces, as in [14]) with a view towards applications in probability, statistical mechanics and statistics.

A survey, with some proofs, of infinite dimensional convexity appears in Fonf–Phelps–Lindenstrauss [38].

The rest of this section will be devoted to brief descriptions of related topics which have been omitted from the body of these notes.

POTENTIAL THEORY

Integral representation theorems play an important role in po-
tential theory, and the Choquet theorem is of considerable use in abstract (or axiomatic) potential theory. Unfortunately, its use in this regard is so deeply imbedded in the subject that it would re-
quire far more time and space then we are willing to spend in order to given an exposition which is even moderately self-contained. What we *can* do is sketch some facts concerning harmonic functions and show how one of the classical integral representation theorems may be viewed as an instance of the Choquet theorem. A much more complete treatment may be found in Becker's book [7].

Let Ω be a bounded, connected, open subset of Euclidean n-
space ($n \geq 2$) and let H be the set of all functions $h \geq 0$ which are harmonic in Ω. Let $E = H - H$ be the linear space generated by H, with the topology of uniform convergence on compact subsets of Ω. Then E is metrizable and H is a closed convex cone which induces a lattice ordering on E. Let x_0 be any point in Ω; then $X = \{h : h \in H, h(x_0) = 1\}$ is a metrizable compact convex base for the cone H. By Choquet's existence and uniqueness theorems, then, to each u in H there exists a unique nonnegative measure μ on the extreme points h of X such that

$$u(x) = \int h(x)d\mu(h) \qquad (x \in \Omega).$$

In view of the characterization (Section 13) of extreme elements of a cone, we see that h lies on an extreme ray of H if and only if $0 \leq u \leq h$, u harmonic, implies $u = \lambda h$ for some $\lambda \geq 0$. Because of this property, the extreme nonnegative harmonic functions are usually referred to as *minimal* harmonic functions.

In order for the above representation theorem to have any significance, of course, one must give a reasonably concrete description of the minimal harmonic functions. For instance, if Ω is the open ball of radius $r > 0$ and center at the origin, and if $x_0 = 0$, then the extreme points come from the Poisson kernel; i.e., a function h in X is extreme if and only if $h = P_y$ for some y with $\|y\| = r$, where

$$P_y(x) = r^{n-2} \frac{r^2 - \|x\|^2}{\|x - y\|^n} \qquad (\|x\| < r).$$

It is easily seen that the map $y \to P_y$ is a homeomorphism from the boundary of the sphere onto ex X, so that the latter is compact (and hence we could have used the Krein-Milman theorem for the existence portion of the above integral representation theorem). The final result, obtained by carrying μ to a measure on the boundary of the sphere, is Herglotz's theorem: If $\mu \in X$, there exists a unique probability measure μ on $\{y : \|y\| = r\}$ such that

(∗) $$u(x) = \int P_y(x) d\mu(y) \qquad (\|x\| < r).$$

The above sketch shows that the Herglotz theorem can be obtained as an application of the Krein-Milman theorem, *provided* one can show that ex X equals the set Y of functions P_y. An elementary proof of this fact for $n = 1$ has been given by F. Holland[43]. An exposition of this and related results may be found in the Lecture Notes of G. Schober[74].

Note that if (∗) holds, then by Milman's theorem, ex X is contained in the closure of Y, and Y is closed. Since rotation of the sphere induces a one-to-one affine map of X onto itself, if one P_y is extreme, they all are. Since ex X is nonempty, we conclude that ex $X = Y$.

POSITIVE DEFINITE FUNCTIONS AND BOCHNER'S THEOREM

A complex valued function f on an Abelian group G is said to be *positive-definite* if

$$\sum_{i,j=1}^{n} \lambda_i \overline{\lambda_j} f(t_i - t_j) \geq 0$$

whenever $t_1, \ldots, t_n$ are elements of G and $\lambda_1, \ldots, \lambda_n$ are complex numbers. It is easily seen that if f is positive definite, then $f(0)$ is real and $|f(t)| \leq f(0)$ for all t in G. If a function f is a *character* of G (i.e., a homomorphism of G into the group of all complex numbers of modulus 1), then f is positive definite. Suppose that G is locally compact and let P be the cone of all continuous positive definite functions on G. Then P can be considered as a subset of the set K of those f in $L^\infty(G)$ which satisfy

$$\iint g(s+t)\overline{g(s)}f(t)\, ds\, dt \geqq 0 \quad (g \in L^1(G)).$$

In the weak* topology, K is closed and has a universal cap, consisting of those f in K with $\|f\| \leq 1$. The nonzero extreme points of this cap are the (essentially) continuous characters χ of G, and it follows that every continuous positive definite function f on G has the form

$$f(t) = \int \chi(t)\, d\mu(\chi)$$

for a nonnegative finite measure μ on the characters. This is a generalization of a classical theorem of Bochner (where G is the real line and each character is of the form $t \to e^{ixt}$ for some real x). Since the extreme points form a closed set, it can actually be proved by the Krein-Milman theorem. It is also possible to use the Stone-Weierstrass theorem to show that μ is uniquely determined by f.

This result has a close connection with group representations, since each continuous positive definite function on G defines, in a canonical way, a continuous unitary representation of G, and the characters correspond to the irreducible representations. The above integral representation essentially shows that every cyclic representation of G (and hence every representation of G) is a "direct integral" of irreducible representations. For further details, see [33] and [58].

It is worthwhile to sketch a simple result which can be used to .show that the extreme points of the set K are the characters. The facts which are left unproved in what follows may be found in [58, §§ 10, 30] (where a "*-algebra" is called a "symmetric ring"). The proof of this result is essentially due to J. L. Kelley and R. L. Vaught[48]. It is applied, of course, to the commutative *-algebra obtained by adjoining the identity to the group algebra $L^1(G)$. A related proof (which doesn't assume continuity of the involution) may be found in Rudin[70].

Suppose that A is a commutative Banach $*$-algebra with identity e and continuous involution $x \to x^*$. Let K be the convex set of all linear functionals f on A which satisfy $f(e) = 1$ and $f(x^*x) \geq 0$ for all x in A. If f is an extreme point of K, then $f(xy) = f(x)f(y)$ for all x, y in A.

PROOF. Any element of A is a linear combination of elements of the form x^*x (consider the polarization identity $x = \frac{1}{4} \sum_{i=1}^{4} \varepsilon_i^{-1} (e + \varepsilon_i x)^*$ $(e + \varepsilon_i x)$, where the ε_i's are the fourth roots of unity); hence we may assume that x is of that form. We may also assume that $\|x^*x\| < 1$. Define the linear functional g on A by $g(y) = f(x^*xy)$. For any y,

$$g(y^*y) = f[(xy)^*(xy)] \geqq 0$$

and

$$(f - g)(y^*y) = f[y^*y(e - x^*x)] = f(y^*y \, z^*z) \geqq 0,$$

since $\|x^*x\| < 1$ implies $e - x^*x = z^*z$, where

$$z^* = z = \sum_{n=0}^{\infty} \binom{1/2}{n} (-x^*x)^n \in A.$$

Thus, $f = g + (f - g)$, where g and $f - g$ are in the cone generated by K. Since f is extreme, we have $g = \lambda f$ for some $\lambda \geq 0$. From $f(e) = 1$ we conclude that $\lambda = g(e)$ and $g(y) = g(e)f(y)$ for all y, i.e., $f(x^*xy) = f(x^*x)f(y)$ for all y which completes the proof.

It is even easier to prove that every multiplicative element of K is extreme. Indeed, if $2f = g + h$, g, h in K, it suffices to prove that $g(x) = 0 = h(x)$ whenever $f(x) = 0$ (since $f(e) = 1 = g(e) = h(e)$ and hence $f = g = h$). But if $f(x) = 0$, then $0 = 2f(x^*)f(x) = 2f(x^*x) = g(x^*x) + h(x^*x)$, so $g(x^*x) = 0 = h(x^*x)$. Furthermore, $|g(x)|^2 \leqq g(x^*x)$, so $g(x) = 0$ and (similarly) $h(x) = 0$.

ANOTHER APPLICATION OF CHOQUET BOUNDARIES AND FUNC-TION ALGEBRAS TO APPROXIMATION THEORY

Bishop's [8] original result concerning a special case of the Choquet theorem in the context of function algebras (and his "peak point" description of the Choquet boundary in this case) was applied to a theorem concerning approximation of continuous complex valued functions on a compact subset of the plane by certain rational functions. We will give the statement of this theorem, and direct the reader to Wermer's monograph [81] for a survey of this and related results.

Let Y be a compact subset of the complex plane, with empty interior. Let A be the subalgebra of $C_c(Y)$ consisting of those continuous complex valued functions on Y which can be uniformly approximated on Y by rational functions which have poles in the complement of Y. It is easily verified that A is uniformly closed, contains the constants, and separates points of Y. Let $B \subset Y$ be the Choquet boundary for A.

THEOREM (Bishop). *The following assertions are equivalent:*

(i) $A = C_c(Y)$.

(ii) $Y \setminus B$ *has two-dimensional Lebesgue measure zero.*

(iii) $B = Y$.

THE SUPPORT OF A MAXIMAL MEASURE

We know that if μ is a maximal probability measure on X, then $\mu(B) = 1$, whenever $\operatorname{ex} X \subset B \subset X$ and B is a Baire set or an F_σ set. This result can be extended to more general sets by means of a theorem of Choquet on abstract capacities. For instance, it is still true if B is a K-Suslin set ($= K$-analytic set) (see [57] for a complete proof) or if B is a K-Borel set [6]; each of these classes of sets contains the Baire sets, and the K-Suslin sets form the largest family. [Since the K-Suslin sets need not be Borel sets, we have to make clear the meaning of "$\mu(B) = 1$." The Choquet-Meyer result can be formulated as follows: If B is a K-Suslin set and $\operatorname{ex} X \subset B$, then for any maximal measure μ there exists an F_σ-set K, with $K \subset B$, such that $\mu(K) = 1$.] The only property of maximal measures used in the proof of either theorem is the fact that $\mu(B) = 1$ if μ is maximal and B is an F_σ set containing $\operatorname{ex} X$.

The K-Borel sets are the simplest to describe: They are the members of the smallest family which contains the compact sets and is closed under countable unions and countable intersections (but not necessarily closed under differences). The family, which need not be a σ-ring, lies between the Baire sets and the Borel sets. It is immediate that Corollary 10.9 on uniqueness can be sharpened to the following form: *If X is a simplex and if ex X is a K-Suslin set, then for each x in X there exists a unique measure μ such that $\mu \sim \varepsilon_x$ and $\mu(\text{ex } X) = 1$ (in the sense described above).*

OTHER EXTENSIONS OF THE KREIN-MILMAN AND MINKOWSKI THEOREMS

As shown in Proposition 13.6, if a cone K in a locally convex space is closed, is locally compact, and contains no line, then K admits a compact base. Thus, local compactness makes it possible to extend the Krein-Milman theorem to (proper) cones, provided we replace "extreme points" by "extreme rays" in the statement of the theorem. (Of course, this is true even under the weaker hypothesis that K is the union of its caps.) What if we drop the hypothesis that K be a cone? Klee [51] has obtained two results in this direction, one which extends Minkowski's theorem on finite dimensional sets and one which extends the Krein-Milman theorem. We first require a definition.

An *extreme ray* of a convex set X is an open half-line $\rho \subset X$ with the property that the open segment $]x, y[$ is contained in ρ whenever $]x, y[\subset X$ and $]x, y[$ intersects ρ. A set is said to be *linearly closed* if its intersection with each line is closed. We let exr X denote the union of the extreme rays of X. Klee's results are the following:

If X is a locally compact closed convex subset of a locally convex space, and if X contains no line, then X is the closed convex hull of ex $X \cup$ exr X.

If X is a linearly closed finite dimensional convex set which contains no line, then X is the convex hull of ex$X \cup$ exr X.

SOME TOPOLOGICAL PROPERTIES OF THE SET OF EXTREME POINTS

As was shown in the Introduction, the set ex X of extreme points of a compact convex subset X of a locally convex space will form a

G_δ set if X is metrizable. In the general case, however, ex X need not even be a Borel set. Nevertheless, Choquet has proved the following results: *If X is a compact convex subset of a locally convex space, then* ex X *is a Baire space in the induced topology.* (Recall that a topological space T is a *Baire* space provided the intersection of any sequence of dense open subsets of T is dense in T.) This result has an interesting application to C^*-algebras, and its proof may be found in [25, p. 355].

R. Haydon [41] has shown that if T is a Polish space (that is, T is a separable metrizable space, complete in some metric compatible with its topology), then there exists a compact convex simplex X such that ex X is homeomorphic to T.

Finally, M. Talagrand [77] has proved a general result which implies that if the set of extreme points of a compact convex set is *K-analytic* (that is, it is the continuous image of a $K_{\sigma\delta}$ subset of a compact Hausdorff space), then it is a Borel set of a special form.

THE POULSEN SIMPLEX.

The fact that the set of extreme points of a simplex need not be closed (as, for instance, is implied by Haydon's theorem quoted above) is contrary to the intuition gained from looking at the finite–dimensional case. Even more counter–intuitive is the remarkable example constructed by E. T. Poulsen [65] of a metrizable simplex X such that ex X is dense in X. It turns out that, up to affine homeomorphism, this simplex – now called the *Poulsen simplex* – is unique. It has other interesting properties, as shown by Lindenstrauss, Olsen and Sternfeld [54]:

THEOREM 16.1 *(i) Uniqueness: There is, up to affine homeomorphism, a unique compact metrizable simplex X with* cl ex$X = X$ *(the Poulsen simplex).*
(ii) Universality: Every compact metrizable simplex is affinely homeomorphic to a face of the Poulsen simplex.
(iii) Homogeneity: For any two extreme points s_1 and s_2 of the Poulsen simplex there is an affine automorphism of the simplex which carries s_1 to s_2. More generally, if F_1 and F_2 are two closed proper faces of the Poulsen simplex and if φ is an affine homeomorphism from F_1 onto F_2 then φ can be extended to an affine automorphism

of the Poulsen simplex.

The Poulsen simplex appears in a surprisingly wide variety of situations. In statistical mechanics, for instance, it arises as the set of states (probability measures) which are invariant under certain natural actions; its (dense) extreme points are the ergodic states. Details may be found in R. Israel [45], Israel-Phelps [46] and D. Ruelle's books [71, 72] as well as in Olsen's survey [59]. A proof is sketched in [38, Sec. 3] of the following result: *Let Z be the set of all integers and $\Omega = \{0,1\}^Z$ be the set of all doubly–infinite sequences of zeros and ones, so (in its product topology) Ω is a compact Hausdorff space. If T is the (continuous) natural shift map of Ω onto iself, then the set X of all T–invariant probability measures on Ω forms the Poulsen simplex, that is, the ergodic measures are dense in X.*

A GEOMETRICAL CHARACTERIZATION OF SIMPLICES

There is an elegant characterization of infinite dimensional simplices which makes no reference to orderings or integral representations, but is purely geometric. We first need the concept of a homothetic image.

DEFINITION 16.2 *Suppose that X is a compact convex set in a topological vector space E. A homothetic image of X is any set of the form $\alpha X + x$, where $\alpha > 0$ and $x \in E$.*

While the intersection of two different homothetic images of X will again be a compact convex set, it will generally bear little resemblance to X. However, if X is any triangle in the plane, a nontrivial intersection of two of its homothetic images will "look" just like X; that is, it will be another homothetic image of X. This observation is the two–dimensional version of the following theorem of Choquet. D. G. Kendall [49] has given a proof which avoids the compactness assumption.

THEOREM 16.3 *A compact convex subset X of a topological vector space E is a simplex if and only if the intersection $(\alpha X + x) \cap (\beta X + y)$ of any two homothetic images of X is either empty, a single point or another homothetic image of X.*

This characterization leads to a simple proof of the following generalization of a theorem of Borovikov [12], who proved it in 1952 for a decreasing sequence of finite dimensional simplices.

THEOREM *The intersection of a downward directed family of compact simplices is a simplex.*

PROOF. Let $\mathcal{X}$ be a directed family of simplices and let $\widehat{X} = \cap_{X \in \mathcal{X}} X$. Let $0 < \alpha, \beta \le 1$, $x, y \in X$ and assume that $(x + \alpha\widehat{X}) \cap (y + \beta\widehat{X}) \ne \emptyset$. Then for every $X \in \mathcal{X}$ there is a z_X and a $0 \le \gamma_X \le 1$ so that $(x + \alpha X) \cap (y + \beta X) = z_X + \gamma_X X$. Choose any $X_0 \in \mathcal{X}$. Then for $X \subset X_0$, z_X belongs to the compact set $x + \alpha X_0 - [0, 1]X_0$. Thus the set $\{z_X, \gamma_X\}_{X \in \mathcal{X}}$ has a cluster point $(\tilde{z}, \tilde{\gamma})$. Since for every $X \in \mathcal{X}$ we have $(x + \alpha\widehat{X}) \cap (y + \beta\widehat{X}) \subset z_X + \gamma_X X$, it follows that $(x + \alpha\widehat{X}) \cap (y + \beta\widehat{X}) \subset \tilde{z} + \tilde{\gamma}\widehat{X}$. Similarly, for every $X \in \mathcal{X}$ we have $z_X + \gamma_X \widehat{X} \subset (x + \alpha X) \cap (y + \beta X)$, which implies that $\tilde{z} + \tilde{\gamma}\widehat{X} \subset (x + \alpha\widehat{X}) \cap (y + \beta\widehat{X})$ and therefore $(x + \alpha\widehat{X}) \cap (y + \beta\widehat{X}) = \tilde{z} + \tilde{\gamma}\widehat{X}$.

Recall that a simplex whose extreme points form a closed set is called a Bauer simplex. D. A. Edwards [32] has shown that any metrizable simplex "is" an intersection of Bauer simplices.

THEOREM 16.4 ([32]) *If X is a metrizable simplex, then there exists a decreasing sequence $X_1 \supset X_2 \supset \cdots \supset X_n \supset \ldots$ of Bauer simplices such that X is affinely homeomorphic to $\cap X_n$.*

FACES AND EDWARDS' SEPARATION THEOREM

The notion of a "face" of a convex set (also called an "extreme" or "extremal subset") is an easily visualized and geometrically appealing concept which appears in many contexts.

DEFINITION 16.5 *A nonempty convex subset F of a convex set X is called a* face *of X provided $y, z \in F$ whenever $x \in F$ and $x = \lambda y + (1 - \lambda)z$ for some $0 < \lambda < 1$.*

For example, the faces of a triangle are its sides, vertices and the entire set. It is clear that, always, a one–point face of X is an extreme point. It is equally clear that the set of points where an affine real–valued continuous function on X attains its maximum is a face.

Faces play an important role in Alfsen's exposition [1] (although they are not explicitly defined) and in that of Asimow and Ellis [2]; in particular, they make use of the notion of a "split" face. Faces appear in the proof of the Krein–Milman theorem, see, for instance [70, p. 74]. Families of face–like sets appear in the proof of *Bauer's maximum principle* [20, Vol. II, p. 102] where the latter is shown to imply the Krein–Milman theorem. Here is the statement:

PROPOSITION 16.6 (BAUER) *A convex upper semicontinuous function on a compact convex set X attains its maximum value (not necessarily uniquely) at an extreme point of X.*

This result is one step in motivating *D. A. Edwards' separation theorem* [31]:

THEOREM 16.7 (EDWARDS) *If f and $-g$ are upper semicontinuous convex functions on the simplex X with $f \leq g$, then there exists a continuous affine function h on X such that $f \leq h \leq g$.*

This can be considered as an extension of the following classical result from real analysis. (See, for instance [34, p. 88].)

Suppose that f and $-g$ are upper semicontinuous functions on the compact Hausdorff space Y such that $f \leq g$. Then there exists a continuous function h on Y such that $f \leq h \leq g$.

To see that this implies Edwards' theorem for *Bauer* simplices, suppose that X is a Bauer simplex and f and $-g$ are upper semicontinuous convex functions on X. Restrict f and g to the compact Hausdorff space $Y = \text{ex } X$, apply the classical result to obtain a separating function h on Y and use the identification of X with the probability measures on Y to realize h as an affine continuous function on X. Bauer's maximum principle shows that $f \leq h \leq g$ on all of X.

UNIQUE REPRESENTATIONS IN THE COMPLEX CASE.

As noted in Section 6, there is no loss in generality in switching from the study of arbitrary compact convex sets to sets of the form $K(M)$, where the latter is the state space of a subspace M of $C(Y)$

(or $C_c(Y)$), for some compact Hausdorff space Y. As before, we assume that M contains the constants and separates the points of Y. In the complex case, the main representation theorem in Section 6 produced, for each $L \in M^*$, a representing complex "boundary" measure μ on Y, but nothing was said about the *norm* of μ. O. Hustad showed [44] that one can, in fact, choose a boundary representing measure μ for which $\|\mu\| = \|L\|$. (An exposition of Hustad's result is contained in [64].) This leads naturally to the question of uniqueness of such norm–preserving representations. To formulate such a result requires the following definition.

DEFINITION 16.8 *A compact convex set K is said to be a* simplexoid *if every closed proper face of K is a simplex.*

Obviously, a compact simplex is a simplexoid, as is an octahedron or the infinite–dimensional version of an octahedron, namely, the unit ball of ℓ_1, in its weak* topology.

THEOREM 16.9 *Let M be a closed subspace of the complex space $C_c(Y)$ of complex continuous functions on the compact metric space Y. Assume that M contains 1 and separates the points of Y. Then each $L \in M^*$ can be represented by a complex Borel measure μ on the Choquet boundary $B(M)$ of M satisfying $\|\mu\| = \|L\|$, and each $L \in M^*$ has a unique such representing measure if and only if the unit ball B_{M^*} of M^* is a simplexoid.*

This result (due to R. Fuhr and the author [39]) can be reformulated and proved for nonmetrizable compact spaces Y; see [39] or the exposition in [64].

CHOQUET THEOREMS FOR NONCOMPACT SETS.

There is a class of convex subsets of Banach spaces for which there exists a strong version of Choquet's existence and uniqueness theorem, namely, the class of all bounded closed convex sets having the *Radon-Nikodým property* (RNP). As the name suggests, the RNP is defined in terms of Radon-Nikodým derivatives of certain vector–valued measures; see Bourgin [14] or Diestel–Uhl [24] for details. A characterization of sets with the RNP (which we will take as the definition since it is closely related to the subject matter of these notes) can be given in terms of the following notion.

DEFINITION 16.10 *A point x in a closed convex subset X of the Banach space E is said to be a* strongly exposed point *if there exists $f \in E^*$ such that $f(x) = \sup\{f(y)\colon\ y \in X\}$ and $\|x_n - x\| \to 0$ whenever $x_n \in X$ and $f(x_n) \to f(x)$.*

DEFINITION 16.11 *A closed convex subset X of a Banach space E is said to have the RNP provided every bounded closed convex subset of X is the closed convex hull of its strongly exposed points.*

Note that since we haven't excluded the case $X = E$, this yields the definition of a *Banach space* with the RNP. Examples of spaces with the RNP: Any reflexive Banach space and any separable dual space (e.g., ℓ_1). The space $L_1[0,1]$ does *not* have the RNP. See [24] for a comprehensive treatment of this property.

The set ex X of extreme points of a bounded closed noncompact subset X of a Banach space (even for a separable X with the RNP) need not be Borel measurable [14], but in the separable case, at least, ex X is always *universally measurable* with respect to the regular Borel measures, that is, for any regular Borel measure μ on X, ex X lies in the completion with respect to μ of the σ-algebra of Borel measurable subsets of X.

THEOREM 16.12 (G. A. EDGAR) *Suppose X is a bounded closed convex and separable subset of the Banach space E and that X has the RNP. For each $x \in X$ there exists a regular Borel probability measure μ on X such that $\mu(\text{ex } X) = 1$ and $f(x) = \int_X f d\mu$ for each $f \in E^*$.*

A uniqueness theorem (due to R. D. Bourgin and G. A. Edgar) and a nonseparable version of this theorem may be found in [14].

References

[1]   Erik M. Alfsen. "Compact Convex Sets and Boundary Integrals," *Ergeb. der Math u. i. Grenzgeb.* Band 57, Springer-Verlag, Berlin-Heidelberg-New York, (1971).

[2]   L. Asimow and A. J. Ellis, "Convexity Theory and its Applications in Functional Analysis," *Academic Press*, (1980).

[3]   Heinz Bauer, "Minimalstellen von Funktionen und Extremalpunkte," *Archiv der Math.* 9 (1958), 389–393.

[4]   Heinz Bauer, "Šilovscher Rand und Dirichletsches Problem," *Ann. Inst. Fourier (Grenoble)* 11 (1961), 89–136.

[5]   Heinz Bauer, "Kennzeichnung kompakter Simplexe mit abgeschlossener Extremalpunktmenge," *Archiv der Math.* 14 (1963) 415–421.

[6]   Heinz Bauer, "Konvexität in topologischen Vektorräumen," lecture notes, University of Hamburg (1963/1964).

[7]   Richard Becker, "Cônes convexes en analyse," Travaux en cours , 59 *Hermann* Paris (1999).

[8]   Errett Bishop, "A minimal boundary for function algebras," *Pacific J. Math.* 9 (1959), 629–642.

[9]   Errett Bishop and Karel de Leeuw, "The representatioan of linear functionals by measures on sets of extreme points," *Ann. Inst. Fourier (Grenoble)* 9 (1959), 305–331.

[10] David Blackwell, "Equivalent comparisons of experiments," *Ann. Math. Stat.* 24 (1953), 265–272.

[11] F. F. Bonsall, "On the representation of points of a convex set," *J. London Math. Soc.* 38 (1963), 332–334.

[12] V. Borovikov, "On the intersection of a sequence of simplices," (Russian) *Uspehi Mat. Nauk* 7 (52) (1952), 179–180.

[13] N. Bourbaki, *Eléménts de Mathématique*, Livre VI, *Intégration*, Ch. 6, "Integration Vectorielle," A. S. I. 1281, Hermann, Paris, (1959).

[14] Richard D. Bourgin, "Geometric Aspects of Convex Sets with the Radon-Nikodým Property," *Lecture Notes in Math. 993* Springer-Verlag (1983).

[15] P. Cartier, J. M. G. Fell and P. A. Meyer, "Comparison des mesures partées par un ensemble convexe compact," *Bull. Soc. Math. Fr.* 92 (1964), 435–445.

[16] Gustave Choquet, "Theory of capacities," *Ann. Inst. Fourier (Grenoble)* 5 (1955), 131–295.

[17] Gustave Choquet, "Existance et unicité des representations intégrales au moyen des points extrémaux dans les cônes convexes," *Séminaire Bourbaki* (Dec. 1956), 139, 15 pp.

[18] Gustave Choquet, "Remarques à propos de le démonstration de l'unicité de P. A. Meyer," *Séminaire Brelot—Choquet—Deny* (Theorie de Potentiel), 6 (1962), No. 8, 13 pp.

[19] Gustave Choquet, "Les cônes convexes faiblement complets dans l'Analyse," *Proc. Intern. Congr. Mathematicians*, Stockholm (1962), 317–330.

[20] Gustave Choquet, "Lectures on Analysis, Vols. I, II, III" *W. A. Benjamin* Eds. J. Marsden, T. Lance and S. Gelbart (1969).

[21] Gustave Choquet, "Représentation intégrale," in "Measure Theory and its Applications," *Lecture Notes in Math. 1033,* Springer–Verlag (1983), 114–143.

[22] Gustave Choquet and Paul-André Meyer, "Existence et unicité des representations intégrals dans les convexes compacts quelconques," *Ann. Inst. Fourier (Grenoble)* 13 (1963), 139–154.

[23] M. M. Day, "Fixed-point theorems for compact convex sets," *Illinois J. Math.* 5 (1961), 585–590.

[24] J. Diestel and J. J. Uhl, JR. "Vector measures" *Math. Surveys 15*, Amer. Math. Soc. (1977).

[25] J. Dixmier, "Les *C**-algèbres et leurs represéntations", Gauthier–Villars, Paris (1964).

[26] K. Donner, "Extension of Positive Operators and Korovkin Theorems," *Lecture Notes in Math.* 904, Springer–Verlag, (1982).

[27] T. Downarowicz, "The Choquet simplex of invariant measures for minimal flows," *Israel J. Math.* 74 (1991), 241–256.

[28] Nelson Dunford and Jacob T. Schwartz, *Linear Operators, Part I*, Interscience, New York–London, (1958).

[29] E. B. Dynkin, "Sufficient statistics and extreme points," *Annals Prob.* 6 (1978), 705–730.

[30] D. A. Edwards, "On the representation of certain functionals by measures on the Choquet boundary," *Ann. Inst. Fourier (Grenoble)* 13 (1963), 111–121.

[31] D. A. Edwards, "Séparation des fonctions réels définies sur un simplexe de Choquet," *C. R. Acad. Sci. Paris* 261 (1965), 2798–2800.

[32] D. A. Edwards, "Systèmes projectifs d'ensembles convexes compacts," *Bull. Soc. Math. France* 103 (1975), 225–240.

[33] R. E. Edwards, *Functional Analysis*, Holt, Rinehart and Winston, New York, (1965).

[34] R. Engelking, *General Topology*, Polska Akad. Nauk. *Monografie Mat.* 60, Warsaw (1977).

[35] Hicham Fakhoury, "Caracterisation des simplexes compacts," *C. R. Acad. Sci. Paris (Ser. A)*, 269 (1969), 21–24.

[36] R. H. Farrell, "Representation of invariant measures," *Illinois J. Math.* 6 (1962), 447–467.

[37] J. Feldman, "Representations of invariant measures," (1963) (dittoed notes, 17 pp.).

[38] V. P. Fonf, J. Lindenstrauss and R. R. Phelps, "Infinite dimensional convexity," *North Holland Handbook on the Geometry of Banach Spaces*, W. B. Johnson and J. Lindenstrauss, editors. (to appear).

[39] R. Fuhr and R.R. Phelps, "Uniqueness of complex representing measures on the Choquet Boundary." *J. Funct. Anal.* 14 (1973), 1–27.

[40] Richard Haydon, "A new proof that every Polish space is the extreme boundary of a simplex." *Bull. London Math. Soc.* 7 (1975), 97–100.

[41] Richard Haydon, "An extreme point criterion for separability of a dual Banach space, and a new proof of a theorem of Corson," *Quarterly J. Math. (Oxford)* 27 (1976), 377–385

[42] Michel Hervé, "Sur les representations intégrales a l'aide des points extrémaux dans un ensemble compact convexe metrizable," *C. R. Acad. Sci.* (Paris) 253 (1961), 366–368.

[43] Finbarr Holland, "The extreme points of a class of functions with positive real part," *Math. Ann.* 202 (1973), 85–87.

[44] O. Hustad, "A norm preserving complex Choquet theorem", *Math. Scand* 29 (1971), 272–278.

[45] R. Israel, "Convexity in the Theory of Lattice Gases", *Princeton Ser. Phys., Princeton University Press*, Princeton, N. J. (1979).

[46] R. Israel and R. R. Phelps, "Some convexity questions arising in statistical mechanics", *Math. Scand.* 54 (1984), 133–156.

[47] J. Kelley, I. Namioka and co-authors, *Linear Topological Spaces*, Van Nostrand, Princeton, N.J., (1963).

[48] J. L. Kelley and R. L. Vaught, "The positive cone in Banach algebras," *Trans. Amer. Math. Soc.* 74 (1953), 44–55.

[49] D. G. Kendall, "Simplexes and vector lattices," *J. Lond. Math. Soc.* 37 (1962), 365–371.

[50] V. L. Klee, Jr., "Some new results on smoothness and rotundity in normed linear spaces," *Math. Ann.* 139 (1959), 295–300.

[51] V. L. Klee, Jr., "Extremal structure of convex sets," *Archiv der Math.* 8 (1957), 234–240.

[52] P. P. Korovkin, "Linear Operators and Approximation Theory," *Hindustan Publishing Co.*, Delhi (1960).

[53] S. S. Kutateladze, "Choquet boundaries in K-spaces," *Russian Math. Surveys* 30: 4 (1975), 115–155.

[54] J. Lindenstrauss, G.H. Olsen and Y. Sternfeld, "The Poulsen simplex," *Ann. Inst. Fourier (Grenoble)* 28 (1978), 91–114.

[55] Georges Lion, "Families résolventes et frontière de Choquet," *C. R. Acad. Sci.* (Paris) 259 (1964), 4460–4462.

[56] L. H. Loomis, "Unique direct integral decompositions on convex sets," *Amer. J. Math.* 94 (1962), 509–526.

[57] P. A. Meyer, *Probability and Potentials*, Blaisdell, New York, (1966).

[58] M. A. Naimark, *Normed Rings*, Noordhoff, Groningen, (1959).

[59] G. H. Olsen, "On simplices and the Poulsen simplex" in *Functional Analysis: Surveys and Recent Results II, Proc. Second Conf. Functional Anal.*, Univ. Paderborn, Paderborn, 1979, Eds. K.-D. Bierstedt and B. Fuchssteiner, *North–Holland Math. Studies 38*, North-Holland, Amsterdam – New York, (1980), pp. 31 - 52.

[60] R. R. Phelps, "Lectures on Choquet's Theorem," *Van Nostrand Math. Studies*, 7 (1966).

[61] R. R. Phelps, "Integral representations for elements of convex sets," *Studies in Functional Analysis*, MAA Studies in Math. 21 (1980), 115–157.

[62] R. R. Phelps, "Convex functions, Monotone Operators and Differentiability," *Lect. Notes in Math. 1364* 2nd Ed., Springer–Verlag (1993).

[63] R. Phelps, "Lektsii o Teoremakh Shoke," *Biblioteka Sbornik Matematika, MIR* Moscow, Transl. E. A. Gorin (1968).

[64] R. R. Phelps, "The Choquet representation in the complex case." *Bull. Amer. Math. Soc.* 83 (1977), 299–312.

[65] E.T. Poulsen, "A simplex with dense extreme boundary," *Ann. Inst. Fourier (Grenoble)* 11 (1961), 83–87.

[66] John Rainwater, "Weak convergence of bounded sequences," *Proc. Amer. Math. Soc.* 14 (1963), 999.

[67] M. Rao, "Measurable selection of representing measures," *Quarterly J. Math.* 22 Ser. 2 (1971), 571–572.

[68] Daniel Ray, "Resolvents, transition functions, and strongly Markovian processes," *Ann. Math.* 70 (1959), 43–72.

[69] Marc Rogalski, "Opérateurs de Lion, projecteurs boréliens et simplexes analytiques," *J. Functional Analysis* 2 (1968), 458–488.

[70] Walter Rudin, "Functional Analysis," *McGraw-Hill*, Second Edition (1991).

[71] D. Ruelle, "Statistical mechanics–rigorous results," *W. A. Benjamin*, New York – Amsterdam (1969).

[72] D. Ruelle, "Thermodynamic formalism," *Encyclopedia Math. Appl.*, Addison–Wesley, Reading, Mass. (1978).

[73] Yu. A. Šaškin, "The Milman-Choquet boundary and approximation theory," *Funct. Anal. Appl.* 1 (1967), 170–171.

[74] Glenn Schober, "Univalent Functions – Selected Topics," *Lect. Notes in Math. 478 , Springer–Verlag* (1975).

[75] V. Strassen, "The existence of probability measures with given marginals," *Ann. Math. Stat.* 36 (1965), 423–439.

[76] Michel Talagrand, "Selection mesurable de mesures maximales simpliciales," *Bull. Sci. Math. (Ser. 2)* 102 (1978), 49–56.

[77] Michel Talagrand, "Sur les convexes compacts dont l'ensemble des points extrémaux est K-analytique," *Bull. Soc. Math. France* 107 (1979), 49–53.

[78] V. S. Varadarajan, "Groups of automorphisms of Borel spaces," *Trans. Amer. Math. Soc.* 109 (1963), 191–220.

[79] G. F. Vincent-Smith, "Measurable selections of simplical maximal measures," *J. London Math. Soc. (Ser. 2)* 7 (1973), 427–428.

[80] Bertram Walsh, "Maximal linear mappings and smooth selection of measures on Choquet boundaries," *Mich. Math. J..* 15 (1968), 51–60.

[81] John Wermer, *Banach Algebras and Analytic Functions, Advances in Math.*, Vol. 1, Fasc. 1, Academic Press, New York and London, (1961).

[82] D. V. Widder, *The Laplace Transform*, Princeton University Press, Princeton, N.J., (1941).

[83] Gerhard Winkler, "Choquet Order and Simplices," *Lecture Notes in Math. 1145*, Springer–Verlag, (1985).

Index of symbols

Index

Lecture Notes in Mathematics

For information about Vols. 1–1570
please contact your bookseller or Springer-Verlag

Vol. 1613: J. Azéma, M. Emery, P. A. Meyer, M. Yor (Eds.), Séminaire de Probabilités XXIX. VI, 326 pages. 1995.

Vol. 1614: A. Koshelev, Regularity Problem for Quasilinear Elliptic and Parabolic Systems. XXI, 255 pages. 1995.

Vol. 1615: D. B. Massey, Le Cycles and Hypersurface Singularities. XI, 131 pages. 1995.

Vol. 1616: I. Moerdijk, Classifying Spaces and Classifying Topoi. VII, 94 pages. 1995.

Vol. 1617: V. Yurinsky, Sums and Gaussian Vectors. XI, 305 pages. 1995.

Vol. 1618: G. Pisier, Similarity Problems and Completely Bounded Maps. VII, 156 pages. 1996.

Vol. 1619: E. Landvogt, A Compactification of the Bruhat-Tits Building. VII, 152 pages. 1996.

Vol. 1620: R. Donagi, B. Dubrovin, E. Frenkel, E. Previato, Integrable Systems and Quantum Groups. Montecatini Terme, 1993. Editors:M. Francaviglia, S. Greco. VIII, 488 pages. 1996.

Vol. 1621: H. Bass, M. V. Otero-Espinar, D. N. Rockmore, C. P. L. Tresser, Cyclic Renormalization and Auto-morphism Groups of Rooted Trees. XXI, 136 pages. 1996.

Vol. 1622: E. D. Farjoun, Cellular Spaces, Null Spaces and Homotopy Localization. XIV, 199 pages. 1996.

Vol. 1623: H.P. Yap, Total Colourings of Graphs. VIII, 131 pages. 1996.

Vol. 1624: V. Brınzanescu, Holomorphic Vector Bundles over Compact Complex Surfaces. X, 170 pages. 1996.

Vol.1625: S. Lang, Topics in Cohomology of Groups. VII, 226 pages. 1996.

Vol. 1626: J. Azéma, M. Emery, M. Yor (Eds.), Séminaire de Probabilités XXX. VIII, 382 pages. 1996.

Vol. 1627: C. Graham, Th. G. Kurtz, S. Méléard, Ph. E. Protter, M. Pulvirenti, D. Talay, Probabilistic Models for Nonlinear Partial Differential Equations. Montecatini Terme, 1995. Editors: D. Talay, L. Tubaro. X, 301 pages. 1996.

Vol. 1628: P.-H. Zieschang, An Algebraic Approach to Association Schemes. XII, 189 pages. 1996.

Vol. 1629: J. D. Moore, Lectures on Seiberg-Witten Invariants. VII, 105 pages. 1996.

Vol. 1630: D. Neuenschwander, Probabilities on the Heisenberg Group: Limit Theorems and Brownian Motion. VIII, 139 pages. 1996.

Vol. 1631: K. Nishioka, Mahler Functions and Transcendence.VIII, 185 pages.1996.

Vol. 1632: A. Kushkuley, Z. Balanov, Geometric Methods in Degree Theory for Equivariant Maps. VII, 136 pages. 1996.

Vol.1633: H. Aikawa, M. Essén, Potential Theory – Selected Topics. IX, 200 pages.1996.

Vol. 1634: J. Xu, Flat Covers of Modules. IX, 161 pages. 1996.

Vol. 1635: E. Hebey, Sobolev Spaces on Riemannian Manifolds. X, 116 pages. 1996.

Vol. 1636: M. A. Marshall, Spaces of Orderings and Abstract Real Spectra. VI, 190 pages. 1996.

Vol. 1637: B. Hunt, The Geometry of some special Arithmetic Quotients. XIII, 332 pages. 1996.

Vol. 1638: P. Vanhaecke, Integrable Systems in the realm of Algebraic Geometry. VIII, 218 pages. 1996.

Vol. 1639: K. Dekimpe, Almost-Bieberbach Groups: Affine and Polynomial Structures. X, 259 pages. 1996.

Vol. 1640: G. Boillat, C. M. Dafermos, P. D. Lax, T. P. Liu, Recent Mathematical Methods in Nonlinear Wave Propagation. Montecatini Terme, 1994. Editor: T. Ruggeri. VII, 142 pages. 1996.

Vol. 1641: P. Abramenko, Twin Buildings and Applications to S-Arithmetic Groups. IX, 123 pages. 1996.

Vol. 1642: M. Puschnigg, Asymptotic Cyclic Cohomology. XXII, 138 pages. 1996.

Vol. 1643: J. Richter-Gebert, Realization Spaces of Polytopes. XI, 187 pages. 1996.

Vol. 1644: A. Adler, S. Ramanan, Moduli of Abelian Varieties. VI, 196 pages. 1996.

Vol. 1645: H. W. Broer, G. B. Huitema, M. B. Sevryuk, Quasi-Periodic Motions in Families of Dynamical Systems. XI, 195 pages. 1996.

Vol. 1646: J.-P. Demailly, T. Peternell, G. Tian, A. N. Tyurin, Transcendental Methods in Algebraic Geometry. Cetraro, 1994. Editors: F. Catanese, C. Ciliberto. VII, 257 pages. 1996.

Vol. 1647: D. Dias, P. Le Barz, Configuration Spaces over Hilbert Schemes and Applications. VII. 143 pages. 1996.

Vol. 1648: R. Dobrushin, P. Groeneboom, M. Ledoux, Lectures on Probability Theory and Statistics. Editor: P. Bernard. VIII, 300 pages. 1996.

Vol. 1649: S. Kumar, G. Laumon, U. Stuhler, Vector Bundles on Curves – New Directions. Cetraro, 1995. Editor: M. S. Narasimhan. VII, 193 pages. 1997.

Vol. 1650: J. Wildeshaus, Realizations of Polylogarithms. XI, 343 pages. 1997.

Vol. 1651: M. Drmota, R. F. Tichy, Sequences, Discrepancies and Applications. XIII, 503 pages. 1997.

Vol. 1652: S. Todorcevic, Topics in Topology. VIII, 153 pages. 1997.

Vol. 1653: R. Benedetti, C. Petronio, Branched Standard Spines of 3-manifolds. VIII, 132 pages. 1997.

Vol. 1654: R. W. Ghrist, P. J. Holmes, M. C. Sullivan, Knots and Links in Three-Dimensional Flows. X, 208 pages. 1997.

Vol. 1655: J. Azéma, M. Emery, M. Yor (Eds.), Séminaire de Probabilités XXXI. VIII, 329 pages. 1997.

Vol. 1656: B. Biais, T. Björk, J. Cvitanic, N. El Karoui, E. Jouini, J. C. Rochet, Financial Mathematics. Bressanone, 1996. Editor: W. J. Runggaldier. VII, 316 pages. 1997.

Vol. 1657: H. Reimann, The semi-simple zeta function of quaternionic Shimura varieties. IX, 143 pages. 1997.

Vol. 1658: A. Pumarino, J. A. Rodrıguez, Coexistence and Persistence of Strange Attractors. VIII, 195 pages. 1997.

Vol. 1659: V. Kozlov, V. Maz'ya, Theory of a Higher-Order Sturm-Liouville Equation. XI, 140 pages. 1997.

Vol. 1660: M. Bardi, M. G. Crandall, L. C. Evans, H. M. Soner, P. E. Souganidis, Viscosity Solutions and Applications. Montecatini Terme, 1995. Editors: I. Capuzzo Dolcetta, P. L. Lions. IX, 259 pages. 1997.

Vol. 1661: A. Tralle, J. Oprea, Symplectic Manifolds with no Kähler Structure. VIII, 207 pages. 1997.

Vol. 1662: J. W. Rutter, Spaces of Homotopy Self-Equivalences – A Survey. IX, 170 pages. 1997.

Vol. 1663: Y. E. Karpeshina; Perturbation Theory for the Schrödinger Operator with a Periodic Potential. VII, 352 pages. 1997.

Vol. 1664: M. Väth, Ideal Spaces. V, 146 pages. 1997.

Vol. 1665: E. Giné, G. R. Grimmett, L. Saloff-Coste, Lectures on Probability Theory and Statistics 1996. Editor: P. Bernard. X, 424 pages, 1997.

Vol. 1666: M. van der Put, M. F. Singer, Galois Theory of Difference Equations. VII, 179 pages. 1997.

Vol. 1667: J. M. F. Castillo, M. González, Three-space Problems in Banach Space Theory. XII, 267 pages. 1997.

Vol. 1668: D. B. Dix, Large-Time Behavior of Solutions of Linear Dispersive Equations. XIV, 203 pages. 1997.

Vol. 1669: U. Kaiser, Link Theory in Manifolds. XIV, 167 pages. 1997.

Vol. 1670: J. W. Neuberger, Sobolev Gradients and Differential Equations. VIII, 150 pages. 1997.

Vol. 1671: S. Bouc, Green Functors and G-sets. VII, 342 pages. 1997.

Vol. 1672: S. Mandal, Projective Modules and Complete Intersections. VIII, 114 pages. 1997.

Vol. 1673: F. D. Grosshans, Algebraic Homogeneous Spaces and Invariant Theory. VI, 148 pages. 1997.

Vol. 1674: G. Klaas, C. R. Leedham-Green, W. Plesken, Linear Pro-p-Groups of Finite Width. VIII, 115 pages. 1997.

Vol. 1675: J. E. Yukich, Probability Theory of Classical Euclidean Optimization Problems. X, 152 pages. 1998.

Vol. 1676: P. Cembranos, J. Mendoza, Banach Spaces of Vector-Valued Functions. VIII, 118 pages. 1997.

Vol. 1677: N. Proskurin, Cubic Metaplectic Forms and Theta Functions. VIII, 196 pages. 1998.

Vol. 1678: O. Krupková, The Geometry of Ordinary Variational Equations. X, 251 pages. 1997.

Vol. 1679: K.-G. Grosse-Erdmann, The Blocking Technique. Weighted Mean Operators and Hardy's Inequality. IX, 114 pages. 1998.

Vol. 1680: K.-Z. Li, F. Oort, Moduli of Supersingular Abelian Varieties. V, 116 pages. 1998.

Vol. 1681: G. J. Wirsching, The Dynamical System Generated by the 3n+1 Function. VII, 158 pages. 1998.

Vol. 1682: H.-D. Alber, Materials with Memory. X, 166 pages. 1998.

Vol. 1683: A. Pomp, The Boundary-Domain Integral Method for Elliptic Systems. XVI, 163 pages. 1998.

Vol. 1684: C. A. Berenstein, P. F. Ebenfelt, S. G. Gindikin, S. Helgason, A. E. Tumanov, Integral Geometry, Radon Transforms and Complex Analysis. Firenze, 1996. Editors: E. Casadio Tarabusi, M. A. Picardello, G. Zampieri. VII, 160 pages. 1998

Vol. 1685: S. König, A. Zimmermann, Derived Equivalences for Group Rings. X, 146 pages. 1998.

Vol. 1686: J. Azéma, M. Émery, M. Ledoux, M. Yor (Eds.), Séminaire de Probabilités XXXII. VI, 440 pages. 1998.

Vol. 1687: F. Bornemann, Homogenization in Time of Singularly Perturbed Mechanical Systems. XII, 156 pages. 1998.

Vol. 1688: S. Assing, W. Schmidt, Continuous Strong Markov Processes in Dimension One. XII, 137 page. 1998.

Vol. 1689: W. Fulton, P. Pragacz, Schubert Varieties and Degeneracy Loci. XI, 148 pages. 1998.

Vol. 1690: M. T. Barlow, D. Nualart, Lectures on Probability Theory and Statistics. Editor: P. Bernard. VIII, 237 pages. 1998.

Vol. 1691: R. Bezrukavnikov, M. Finkelberg, V. Schechtman, Factorizable Sheaves and Quantum Groups. X, 282 pages. 1998.

Vol. 1692: T. M. W. Eyre, Quantum Stochastic Calculus and Representations of Lie Superalgebras. IX, 138 pages. 1998.

Vol. 1694: A. Braides, Approximation of Free-Discontinuity Problems. XI, 149 pages. 1998.

Vol. 1695: D. J. Hartfiel, Markov Set-Chains. VIII, 131 pages. 1998.

Vol. 1696: E. Bouscaren (Ed.): Model Theory and Algebraic Geometry. XV, 211 pages. 1998.

Vol. 1697: B. Cockburn, C. Johnson, C.-W. Shu, E. Tadmor, Advanced Numerical Approximation of Nonlinear Hyperbolic Equations. Cetraro, Italy, 1997. Editor: A. Quarteroni. VII, 390 pages. 1998.

Vol. 1698: M. Bhattacharjee, D. Macpherson, R. G. Möller, P. Neumann, Notes on Infinite Permutation Groups. XI, 202 pages. 1998.

Vol. 1699: A. Inoue, Tomita-Takesaki Theory in Algebras of Unbounded Operators. VIII, 241 pages. 1998.

Vol. 1700: W. A. Woyczyński, Burgers-KPZ Turbulence, XI, 318 pages. 1998. ´

Vol. 1701: Ti-Jun Xiao, J. Liang, The Cauchy Problem of Higher Order Abstract Differential Equations, XII, 302 pages. 1998.

Vol. 1702: J. Ma, J. Yong, Forward-Backward Stochastic Differential Equations and Their Applications. XIII, 270 pages. 1999.

Vol. 1703: R. M. Dudley, R. Norvaiša, Differentiability of Six Operators on Nonsmooth Functions and p-Variation. VIII, 272 pages. 1999.

Vol. 1704: H. Tamanoi, Elliptic Genera and Vertex Operator Super-Algebras. VI, 390 pages. 1999.

Vol. 1705: I. Nikolaev, E. Zhuzhoma, Flows in 2-dimensional Manifolds. XIX, 294 pages. 1999.

Vol. 1706: S. Yu. Pilyugin, Shadowing in Dynamical Systems. XVII, 271 pages. 1999.

Vol. 1707: R. Pytlak, Numerical Methods for Optimal Control Problems with State Constraints. XV, 215 pages. 1999.

Vol. 1708: K. Zuo, Representations of Fundamental Groups of Algebraic Varieties. VII, 139 pages. 1999.

Vol. 1709: J. Azéma, M. Émery, M. Ledoux, M. Yor (Eds), Séminaire de Probabilités XXXIII. VIII, 418 pages. 1999.

Vol. 1710: M. Koecher, The Minnesota Notes on Jordan Algebras and Their Applications. IX, 173 pages. 1999.

Vol. 1711: W. Ricker, Operator Algebras Generated by Commuting Projections: A Vector Measure Approach. XVII, 159 pages. 1999.

Vol. 1712: N. Schwartz, J. J. Madden, Semi-algebraic Function Rings and Reflectors of Partially Ordered Rings. XI, 279 pages. 1999.

Vol. 1713: F. Bethuel, G. Huisken, S. Müller, K. Steffen, Calculus of Variations and Geometric Evolution Problems. Cetraro, 1996. Editors: S. Hildebrandt, M. Struwe. VII, 293 pages. 1999.

Vol. 1714: O. Diekmann, R. Durrett, K. P. Hadeler, P. K. Maini, H. L. Smith, Mathematics Inspired by Biology. Martina Franca, 1997. Editors: V. Capasso, O. Diekmann. VII, 268 pages. 1999.

Recent Reprints and New Editions

4. Lecture Notes are printed by photo-offset from the master-copy delivered in camera-ready form by the authors. Springer-Verlag provides technical instructions for the preparation of manuscripts. Macro packages in T_EX, L^AT_EX2e, $L^AT_EX2.09$ are available from Springer's web-pages at

http://www.springer.de/math/authors/b-tex.html.

Careful preparation of the manuscripts will help keep production time short and ensure satisfactory appearance of the finished book.

The actual production of a Lecture Notes volume takes approximately 12 weeks.

5. Authors receive a total of 50 free copies of their volume, but no royalties. They are entitled to a discount of 33.3% on the price of Springer books purchase for their personal use, if ordering directly from Springer-Verlag.

Commitment to publish is made by letter of intent rather than by signing a formal contract. Springer-Verlag secures the copyright for each volume. Authors are free to reuse material contained in their LNM volumes in later publications: A brief written (or e-mail) request for formal permission is sufficient.

Addresses:

Professor J.-M. Morel
CMLA, Ecole Normale Supérieure de Cachan
61 Avenue du Président Wilson
94235 Cachan Cedex France
E-mail: Jean-Michel.Morel@cmla.ens-cachan.fr

Professor B. Teissier
Université Paris 7
UFR de Mathématiques
Equipe Géométrie et Dynamique
Case 7012
2 place Jussieu
75251 Paris Cedex 05
E-mail: Teissier@ens.fr

Professor F. Takens, Mathematisch Instituut,
Rijksuniversiteit Groningen, Postbus 800,
9700 AV Groningen, The Netherlands
E-mail: F.Takens@math.rug.nl

Springer-Verlag, Mathematics Editorial, Tiergartenstr. 17
D-69121 Heidelberg, Germany
Tel.: *49 (6221) 487-701
Fax: *49 (6221) 487-355
E-mail: lnm@Springer.de